Lecture Notes in Computer Science   10651

Commenced Publication in 1973
Founding and Former Series Editors:
Gerhard Goos, Juris Hartmanis, and Jan van Leeuwen

More information about this series at http://www.springer.com/series/7409

Sergio de Cesare · Ulrich Frank (Eds.)

Advances in Conceptual Modeling

ER 2017 Workshops AHA, MoBiD,
MREBA, OntoCom, and QMMQ
Valencia, Spain, November 6–9, 2017
Proceedings

Springer

Editors
Sergio de Cesare
University of Westminster
London
UK

Ulrich Frank ⓘ
University of Duisburg-Essen
Essen
Germany

ISSN 0302-9743   ISSN 1611-3349   (electronic)
Lecture Notes in Computer Science
ISBN 978-3-319-70624-5   ISBN 978-3-319-70625-2   (eBook)
https://doi.org/10.1007/978-3-319-70625-2

Library of Congress Control Number: 2017957862

LNCS Sublibrary: SL3 – Information Systems and Applications, incl. Internet/Web, and HCI

Printed on acid-free paper

This Springer imprint is published by Springer Nature
The registered company is Springer International Publishing AG
The registered company address is: Gewerbestrasse 11, 6330 Cham, Switzerland

Preface

At many conferences, workshops play an important role. On the one hand, they support developing new, promising topics. On the other hand, they serve as an inspiring forum for meeting researchers that work in the same field, and, hence, contribute to creating scientific communities. For long, workshops have been an important part of the ER conference series. We are very glad that this year's workshops impressively carried on that strong tradition. The workshops' topics covered two main categories.

New application areas enabled by advanced information technologies demand for dedicated modeling approaches that account for the peculiarities of new domains by providing specific concepts and methods. Corresponding approaches to conceptual modeling do not only facilitate the appropriate design and use of systems that play a key role in the digital transformation, but also contribute to technology acceptance and user empowerment. The second category comprises foundational aspects of research on conceptual modeling. These include the semantics of language concepts, assumptions we use to justify modeling decisions, domain-specific theories, and criteria we use to assess the quality of models.

Three workshops address the first category. The Third International Workshop on Modeling for Ambient Assistance and Healthy Ageing (AHA), organized by Heinrich Mayr, Ulrich Frank, and J. Palazzo M. de Oliveira was motivated by the ever-increasing demand for supporting the elderly in managing their lives. Related research and development is driven into various directions among which the fields of "active and assisted living (AAL)" and "healthy ageing (HA)" are rather prominent. For innovation to work, the stakeholders of future assistance systems have to be involved in order to offer them comprehensible representations of possible solutions that enable them to express their concerns and demands. Therefore, the realization of advanced systems to support AAL and HA recommends the design and use of powerful models. The Sixth International Workshop on Modeling and Management of Big Data (MoBiD) resulted from three workshop proposals that were all focused on modeling approaches to analyze and manage big data, which is one more indicator of the relevance of this field. The workshop was organized by Thomas Bäck, David Gil, Nicholas Multari, Jesús Peral, Heike Trautmann, Juan Trujillo, Il-Yeol Song, and Gottfried Vossen. The MoBiD workshop was dedicated to presenting and discussing new approaches to conceptualizing large amounts of data in order to promote managing and exploiting them. The Fourth International Workshop on Conceptual Modeling in Requirements and Business Analysis (MREBA) was organized by Renata Guizzardi, Eric-Oluf Svee, and Jelena Zdravkovic. Requirements engineering has used conceptual models for long. However, often, corresponding methods are generic. In order to support the analysis and design of business information systems, it seems promising to enrich generic methods with concepts and methods for business analysis. That would include accounting for competitiveness, for costs benefit ratios and further, more specific business performance indicators. The MREBA workshop provides a forum for researchers who work on specific methods for business analysis and on the integration of those methods with approaches to requirements engineering.

The second category comprises two workshops. The Fifth International Workshop on Ontologies and Conceptual Modeling, organized by Frederik Gailly, Giancarlo Guizzardi, Mark Lycett, Chris Partridge, and Michael Verdonck, was aimed at investigating the application of foundational ontologies to the field of conceptual modeling. Ontologies not only promise to provide a consistent foundation for modeling languages, they also propose principles that guide the conceptualization of domains of interest. The workshop is also supposed to provide a forum for the discussion of case studies that demonstrate the application of ontologies. Finally, the 4th Workshop on Quality of Models and Models of Quality (QMMQ), organized by Samira Si-Said, Ignacio Panach, and Pnina Soffer was dedicated to a topic that has been at the core of research on conceptual modeling for long. While it is widely accepted that conceptual models are suited to promoting the quality of software systems, it is evident at the same time that this effect is not independent of the quality of models themselves. Only if we have an elaborate idea of how to assess the quality of models, it is possible to convincingly compare models and modeling approaches. The QMMQ workshop aimed at fostering discussions not only on the quality of conceptual models, but also on methods and tools to promote model quality.

We would like to thank the workshop organizers who prepared and organized inspiring events that clearly contributed to making ER 2017 a special event. We are also indebted to the numerous reviewers who devoted their time and expertise to ensure the quality of the workshops. Finally, we express our gratitude to the general chairs of the conference. It has been a pleasure to work with them.

October 2017   Sergio de Cesare
                                                                          Ulrich Frank

ER 2017 Conference Organization

General Conference Chair

Oscar Pastor — Universitat Politècnica de València, Spain

Organizing Co-chairs

Ignacio Panach — Universitat Politècnica de València, Spain
Victoria Torres — Universitat Politècnica de València, Spain

Industry Chairs

Elena Kornyshova — Centre d'études et de recherché en informatique et communications, France
Vicente Pelechano — Universitat Politècnica de València, Spain
Tanja Vos — Universitat Politècnica de València, Spain

Workshop Co-chairs

Sergio de Cesare — University of Westminster, UK
Ulrich Frank — University of Duisburg-Essen, Germany

Panel Co-chairs

Flavia Santoro — UNIRIO, Brazil
Juan Carlos Trujillo — University of Alicante, Spain
Gerti Kappel — Tu Wien, Austria

PhD Symposium Co-chairs

Dimitris Karagiannis — University of Vienna, Austria
Stephen Liddle — Brigham Young University, USA
Sourav S. Bhowmick — NTU Singapore

Program Committee Co-chairs

Heinrich C. Mayr — AAU, Austria
Hui Ma — VUW, New Zealand
Giancarlo Guizzardi — UFES, Brazil

Tutorial Co-chairs

Iris Reinhartz-Berger   University of Haifa, Israel
Karen Davis   University of Cincinnati, USA

Poster/Tool Demonstration Co-chairs

Sergio España   Universiteit Utrecht, The Netherlands
Cristina Cabanillas   Vienna University of Economics and Business, Austria

SCME Co-chairs

Xavier Franch   UPC, Spain
Monique Snoeck   KU Leuven, Belgium

Publicity Co-chairs

Selmin Nurcan   Sorbone, France
Renata Guizzardi   UFES, Brazil
Moon Kun Lee   Chonbuk National University, Korea

Liaison to Steering Committee

Antoni Olivé   Universitat Politècnica de València, Spain

ER 2017 Workshops Organization

AHA 2017 Co-chairs

Ulrich Frank   University of Duisburg-Essen, Germany
Heinrich C. Mayr   · AAU, Austria
J. Palazzo M. de Oliveira   UFRGS Porto Alegre, Brazil

MoBiD 2017 Co-chairs

David Gil   University of Alicante, Spain
Juan Trujillo   University of Alicante, Spain
Gottfried Vossen   ERCIS, University of Münster, Germany
Thomas Bäck   Leiden University and Divis Intelligent Solutions
                                 GmbH, The Netherlands
Heike Trautmann   ERCIS, University of Münster, Germany
Nicholas Multari   Pacific Northwest National Laboratory, USA
Jesús Peral   University of Alicante, Spain
Il-Yeol Song   Drexel University, USA

MREBA 2017 Co-chairs

Renata Guizzardi   Universidade Federal do Espírito Santo (UFES), Brazil
Eric-Oluf Svee   Stockholm University, Sweden
Jelena Zdravkovic   Stockholm University, Sweden

OntoCom 2017 Co-chairs

Frederik Gailly   Ghent University, Belgium
Chris Partridge   BORO Solutions Ltd., UK
Giancarlo Guizzardi   Federal University of Espirito Santo, Brazil
Mark Lycett   Royal Holloway, University of London, UK
Michael Verdonck   Ghent University, Belgium

QMMQ 2017 Co-chairs

Samira Si-said Cherfi   Conservatoire National des Arts et Métiers, France
Ignacio Panach   Universitat de València, Spain
Pnina Soffer   University of Haifa, Israel

Organization

AHA 2017 Program Committee

Abdelhamid Bouchachia	Bournemouth University, UK
Vadim Ermolayev	Zaporozhye National University, Ukraine
Hans-Georg Fill	Universität Wien, Austria
Sven Hartmann	Universität Clausthal, Germany
Dimitris Karagiannis	Universität Wien, Austria
Gerhard Lakemeyer	RWTH Aachen, Germany
Ivan Lee	University of South Australia, Adelaide
Stephen Liddle	Kevin and Debra Rollins Center for e-Business, USA
Elisabeth Métais	Laboratory CEDRIC, Paris, France
Judith Michael	Alpen-Adria-Universität Klagenfurt, Austria
Oscar Pastor	University of Valencia, Spain
Wolfgang Reisig	Humboldt-Universität zu Berlin, Germany
Pietro Siciliano	IMM-CNR, Italy
Elmar Sinz	Universität Bamberg, Germany
Vladimir Shekhovtsov	National Technical University KhPI, Kharkiv, Ukraine
Markus Stumptner	University of South Australia, Australia
Bernhard Thalheim	Universität Kiel, Germany
Rajesh Vasa	Deakin University, Australia
Benkt Wangler	Stockholm University, Sweden
Shuai Zhang	University of Ulster, UK

MoBiD 2017 Program Committee

Yuan An	Drexel University, Philadelphia, USA
Marie-Aude Aufaure	Ecole Centrale Paris, France
Rafael Berlanga	Universitat Jaume I, Spain
Sandro Bimonte	Irstea, France
Michael Blaha	Yahoo Inc., USA
Hendrik Blockeel	KU Leuven, Belgium
Gennaro Cordasco	Universit di Salerno, Italy
Dickson Chiu	University of Hong Kong, SAR China
Alfredo Cuzzocrea	University of Calabria, Italy
Stuart Dillon	The University of Waikato Management School, New Zealand
Gill Dobbie	University of Auckland, New Zealand
Michael Emmerich	Leiden University, The Netherlands
Jose Luis Fernández-Alemán	University of Murcia, Spain
Johann-Christoph Freytag	Humboldt University Berlin, Germany

Pedro Furtado	Universidade de Coimbra, Portugal
Matteo Golfarelli	University of Bologna, Italy
Bodo Hüsemann	Informationsfabrik GmbH, Münster, Germany
H.V. Jagadish	University of Michigan, USA
John R. Johnson	Pacific Northwest National Lab, USA
Magnus Johnsson	University of Lund, Sweden
Nectarios Koziris	Technical University of Athens, Greece
Karsten Kraume	arvato CRM, Germany
Jiexun Li	Drexel University, Philadelphia, USA
Stephen W. Liddle	BYU - Marriott School, USA
Alexander Löser	Beuth University of Applied Sciences Berlin, Germany
Antoni Olivé	Polytechnic University of Catalonia, Spain
M. Tamer Özsu	University of Waterloo, Canada
Jeff Parsons	Memorial University of Newfoundland, Canada
Oscar Pastor	Polytechnic University of Valencia, Spain
Mario Piattini	University of Castilla-La Mancha, Spain
Nicolas Prat	ESSEC Business School, France
Sudha Ram	University of Arizona, USA
Carlos Rivero	University of Idaho, USA
Colette Roland	University of Paris 1-Pantheon Sorbonne, France
Pablo Sánchez	University of Cantabria, Spain
Dennis Shasha	New York University, NY, USA
Meinolf Sellmann	General Electric, USA
Keng Siau	University of Nebraska-Lincoln, USA
Alkis Simitsis	Hewlett-Packard Co., California, USA
Jeffrey Ullman	Stanford University, USA
Alejandro Vaisman	Université Libre de Bruxelles, Belgium
Panos Vassiliadis	University of Ioannina, Greece

MREBA 2017 Program Committee

Okhaide Akhigbe	University of Ottawa, Canada
Claudia Cappelli	NP2TEC/Universidade Federal do Estado do Rio de Janeiro, Brazil
Fabiano Dalpiaz	Utrecht University, Netherlands
Marcela Ruiz	Universitat Politècnica de València, Spain
Aditya Ghose	University of Wollongong, Australia
Paul Johannesson	KTH Royal Institute of Technology, Sweden
Sotirios Liaskos	York University, Canada
Lin Liu	Tsinghua University, China
Lidia Lopez	Universitat Politècnica de Catalunya, Spain
Pericles Loucopoulos	University of Manchester, UK
Joshua Nwokeji	Gannon University, USA
Andreas Opdahl	University of Bergen, Norway
Anna Perini	Fondazione Bruno Kessler, Italy
Jolita Ralyté	University of Geneva, Switzerland

Kevin Ryan   University of Limerick, Ireland
Junko Shirogane   Tokyo Woman's Christian University, Japan
Samira Si-Said Cherfi   Conservatoire National des Arts et Métiers, France
Vitor Souza   Universidade Federal do Espírito Santo, Brazil
Sam Supakkul   Sabre Travel Network, USA
Lucineia Thom   Universidade Federal do Rio Grande do Sul, Brazil

OntoCom 2017 Program Committee

Organizing Committee

Frederik Gailly   Ghent University, Belgium
Giancarlo Guizzardi   Federal University of Espirito Santo, Brazil
Mark Lycett   Royal Holloway, University of London, UK
Chris Partridge   BORO Solutions Ltd., UK
Michael Verdonck   Ghent University, Belgium
Lucineia Thom   Universidade Federal do Rio Grande do Sul, Brazil

Steering Committee

Oscar Pastor   Polytechnic University of Valencia, Spain
Sergio de Cesare   University of Westminster, UK

QMMQ 2017 Program Committee

Jacky Akoka   CNAM, France
Said Assar   Telecom Ecole de Management, France
Marko Bajec   University of Ljubljana, Slovenia
Cristina Cachero   Universidad de Alicante, Spain
Isabelle Comyn-Wattiau   CNAM-ESSEC, France
Sophie Dupuy-Chessa   Grenoble University, France
Cesar Gonzalez-Perez   Spanish National Research Council, Institute
                                     of Heritage Sciences, Spain
Roberto E. Lopez-Herrejon   Johannes Kepler Universität, Austria
Raimundas Matulevicius   University of Tartu, Estonia
Jeffrey Parsons   University of Newfoundland, Canada
Jolita Ralyte   University of Geneva, Switzerland
Sudha Ram   University of Arizona, USA
Camille Salinesi   CRI, Université de Paris1 Panthéon-Sorbonne
Guttorm Sindre   Norwegian University of Science and Technology,
                                     Norway
Pnina Soffer   IS Department, University of Haifa, Israel

Contents

OntoCom 2017 - 5th International Workshop on Ontologies and Conceptual Modeling

QMMQ 2017 - 4th Workshop on Quality of Models and Models of Quality

AHA 2017 - 3rd International Workshop on Modeling for Ambient Assistance and Healthy Ageing

Preface

Heinrich C. Mayr[1], Ulrich Frank[2], and J. Palazzo M. de Oliveira[3]

[1] Alpen-Adria-Universität Klagenfurt, Klagenfurt, Austria
heinrich.mayr@aau.at
[2] Universität Essen-Duisburg, Duisburg, Germany
ulrich.frank@uni-duisburg-essen.de
[3] UFRGS Porto Alegre, Porto Alegre, Brazil
palazzo@inf.ufrgs.br

"Health, demographic change and wellbeing" is a challenge of the entire world. This becomes evident when looking at current demographic statistics. Related research and development is driven into various directions amongst which the fields of "Active and Assisted Living (AAL)" and "Healthy Ageing (HA)" are rather prominent. The design of innovative and beneficial IT solutions in these domains recommends "thinking out of the box", i.e. looking beyond current ways of living in the older age. For innovation to work, the stakeholders of future assistance systems have to be involved in order to offer them comprehensible representations of possible solutions that enable them to express their concerns and demands. Therefore, the realization of advanced systems to support AAL and HA recommends the design and use of powerful models.

The AHA workshop series is to reveal and promote the existing and potential contributions, which can be made by the modeling community. A particular emphasis is on modeling within the context of designing and developing systems for assisting humans in their everyday live and in healthy ageing. The workshop combines reports on up-to-date research and impulse talks conveying the state-of-the art of conceptual modeling in the AHA domain as well as challenges for future research.

Although AHA2017 is part of the International Conference on Conceptual Modeling, the workshop also aims at attracting researchers, who use other modeling approaches in order to discuss possible overlaps.

All submitted papers have been peer reviewed by at least three members of the program committee. This chapter contains those papers, which have been accepted by the program committee, and carefully revised following the reviewers' comments.

Junsup Song, Maryam Rahmani and Moonkun Lee discuss in their paper *Behavior Ontology to Model Collective Behavior of Emergency Medical Systems* a practical method to model collective behaviors based on a domain ontology and on regular expressions. This is to overcome structural representation limitations that result from 'state explosion'. The practicability of the approach is demonstrated on the basis of two example domains: Emergency Medical Service and Health Care Service.

In the paper *Data Modelling for Dynamic Monitoring of Vital Signs: Challenges and Perspectives* Natalija Kozmina, Emil Syundyukov, and Aleksejs Kozmins focus on data modeling as a means for real-time and complex historical analysis of patients' data. A conceptual model of the domain of knee joint dynamics monitoring is presented, which is exploited by a mobile application for calculations, data analysis and visualization.

The paper *Towards Care Systems Using Model-driven Adaptation and Monitoring of Autonomous Multi-clouds* by Andreea Buga, Sorana Tania Nemeş, and Klaus-Dieter Schewe addresses the application of autonomous multi-clouds in the case of a robotic care system. An Abstract State Machines-based conceptual model is presented which is to specify the middleware architecture and the concrete interaction with the multiple clouds.

Agnes Koschmider addresses in her paper *Clustering Event Traces by Behavioral Similarity* the comparison of event traces in relation to time, duration and exogenous factors by use of a clustering technique. Two algorithms for behavioral morphing and assignment are presented, that support the detection and analysis of behavioral deviations. The approach is based on smart home sensors which transmit event streams which must be splitted up according to entities resulting in an entity-centric trace.

We thank all authors for submitting their excellent scholarly work to AHA2017 as well as the members of our renowned program committee for their careful and intensive collaboration. Likewise we express our gratitude to the ER2017 workshop chairmen for their engagement in supporting our workshop and compiling this volume.

Behavior Ontology to Model Collective Behavior of Emergency Medical Systems

Junsup Song, Maryam Rahmani, and Moonkun Lee[(⊠)]

Chonbuk National University, 567 Beakje-dearo Deokjin-gu,
Jeonju-si Jeonju 54896, Republic of Korea
moonkun@jbnu.ac.kr

Abstract. It is very important to understand system *behavior*s in collective pattern for each knowledge domain. However, there are structural limitations to represent collective behaviors due to the size of system components and the complexity of their interactions, causing the state explosion problem. Further composition with other systems is mostly impractical due to exponential growth of their size and complexity. This paper presents a practical method to model the collective behaviors, based on a new concept of domain engineering: *behavior ontology*. Two domains are selected to demonstrate the method: *Emergency Medical Service* (EMS) and *Health Care Service* (HCS) systems. The examples show that the method is very effective and efficient to construct a hierarchy of collective behaviors in a lattice and that the composition of two collective behaviors is systematically performed by the composition operation of two lattices. The method can be one of the most innovative approaches in representing system behaviors in collective pattern, as well as in minimization of system states to reduce system complexity. For implementation, a prototype tool, called PRISM, has been developed on ADOxx Meta-Modelling Platform.

Keywords: Collective behavior · Behavior ontology · PRISM · ADOxx

1 Introduction

There are strong needs to represent system *behavior*s for each knowledge domain in some collective patterns. However, the needs cannot be easily satisfied due to the structural limitations caused by the size of system components and the complexity of their interactions, as well as their composition, causing *state explosion* [1].

In order to overcome these limitations, this paper presents a method to model the collective behaviors of systems, based on a concept of *behavior ontology* [2]. The research in this paper extended the previous research [2] to other domains with composition for larger and more complex scalability. The approach is as follows:

(1) Firstly, a class hierarchy of a domain is constructed based on active ontology, where all the actors of the domain and their interactions are defined as classes and relations, respectively.

(2) Secondly, each collective behavior of the domain is defined in *regular expression*, where each behavior is defined as a sequence of interactions among actors. The

© Springer International Publishing AG 2017
S. de Cesare and U. Frank (Eds.): ER 2017 Workshops, LNCS 10651, pp. 5–15, 2017.
https://doi.org/10.1007/978-3-319-70625-2_1

behaviors will be presented in a hierarchical order based on their inclusion relations, forming a special lattice, called *n:2-Lattice* [3].

(3) Thirdly, each behavior is quantifiably abstracted with a notion of cardinality and capacity for actors in behavior. This notion will be used to select appropriate behaviors and their relation from the lattice and to make quantitatively equivalent composition with other lattices.

(4) Fourthly, the abstract behavior lattice, *Abstract n:2-Lattice*, is constructed.

(5) Finally, two abstract behavior lattices can be composed with respect to same cardinality and capacity for common actors between two lattices. It implies quantitatively equivalent composition of two types of collective behaviors.

In order to demonstrate the feasibility of the approach, *Emergency Medical Service* (EMS) and *Health Care Service* (HCS) systems are presented for each steps. The examples show that the method is very effective and efficient to construct a hierarchy of collective behaviors in a lattice and that the composition of two collective behaviors is systematically performed by the composition operation of two lattices. The method can be considered to be one of the most innovative approaches in representing system behaviors in collective pattern, as well as in minimization of system states to reduce system complexity. Further, a tool, called PRISM, has been developed on ADOxx Meta-Modelling Platform in order to demonstrate its feasibility.

This paper is organized as follows. Section 2 presents the approach in steps with EMS and HCS. Section 3 analyzes the approach and compares it with other approaches. Section 4 presents the architecture of PRISM. Finally, conclusions and future research will be made in Sect. 5.

2   Approach

2.1   Step 1: Active Ontology

The first step is to design Active Ontology for EMS and HCS services. Active ontology consists of classes and subclasses in a domain, including their interactions.

A. Emergency Medical Service (EMS)

EMS service contains four classes: *Ambulance* (A), *Patient* (P), and *Place* (PL). Note that Place contains *Location* (L) and *Hospital* (H) as subclasses. Similarly, Hospital includes *Bill* (B) as subclass, too. The system's elements represent in the left side of Fig. 1, as follows:

- *Actors*: There are 4 different kinds of actors:
    ① Patient: Person to be transported.
    ② Ambulance: Actor to deliver object.
    ③ Location: Place for Patient to be delivered from.
    ④ Hospital: Place for Patient to be delivered to.

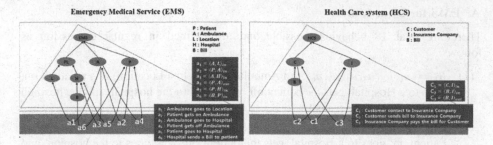

Fig. 1. Active Ontology for Emergency Medical Service (EMS) and Health Care system (HCS)

- *Interactions*: There are 6 kinds of interactions:

  ① $a_1 = <A, L>$: Ambulance goes to Location.
  ② $a_2 = <P, A>$: Patient gets on Ambulance.
  ③ $a_3 = <A, H>$: Ambulance goes to Hospital.
  ④ $a_4 = <A, P>$: Patient gets off Ambulance.
  ⑤ $a_5 = <P, H>$: Patient goes to Hospital.
  ⑥ $a_6 = <H, B>$: Hospital sends Bill to Patient.

B. Health Care Service (HCS)

This system contains two classes: *Customer* (C) and *Insurance Company* (I). And Customer has one subclass: *Bill* (B). The right side of Fig. 1 shows the relationship between classes and subclass, as follows:

- *Actors*: There are 2 different kinds of actors:

  ① Customer: Person to be insured.
  ② Insurance Company: Company to insure Customer.

- *Actions*: There are 3 kinds of actions:

  ① $c_1 = <C, I>$: Customer contacts to Insurance Company.
  ② $c_2 = <C, B>$: Customer sends a Bill to Insurance Company.
  ③ $c_3 = <I, B>$: Insurance Company pays the Bill for Customer.

2.2   Step 2: Regular Behaviors

In this step, each collective behavior is defined as a sequence of interactions from step 1. In order to quantify the behaviors, all behaviors are divided into two kinds of behaviors: the one with one main actor and the other with more than one actor. In the other words, there are different views by different actors. For example, in EMS there are four kind of actors, represented as B(L, A, H, P). Then, there are two behaviors, represented as B(n, 1, n, n) for 1 Ambulance and B(n, n, n, n) for n Ambulances. Similarly to B(C, I) for HCS, B(1, n) for 1 Customer and B(n, n) for n Customer.

A. EMS for B (n, 1, n, n)

There are total 18 behaviors possible and are defined in regular expression as follows:

(1) $B_1 = \langle a_1, a_2, a_3, a_4, a_5, a_6 \rangle$: An Ambulance goes to a Location, gets a Patient on, goes to a Hospital, gets the patient off, who goes to the hospital, and the hospital sends a Bill to the patient.

(2) $B_2 = \langle a_1, a_2, a_3, a_4, a_5, \langle a_6 \rangle^+ \rangle$: An Ambulance goes to a Location, gets a Patient on, goes to a Hospital, gets the patient off, who goes to the hospital, and the hospital sends a number of Bills to the patient.

(3) $B_3 = \langle a_1, a_2, a_3, a_4, a_5, a_6^+ \rangle$: A repeating behavior of B_1.

(4) $B_4 = \langle a_1, a_2, a_3, a_4, a_5, \langle a_6 \rangle^+ \rangle^+$: A repeating behavior of B_2.

(5) $B_5 = \langle a_1, \langle a_2 \rangle^+, a_3, \langle a_4, a_5, a_6 \rangle^+ \rangle^+$: An Ambulance goes to a Location, gets Patients on, goes to a Hospital, gets the patients off, who go to the hospital, and the hospital sends a Bill to each patient. And it repeats itself.

(6) $B_6 = \left\langle a_1, \langle a_2 \rangle^+, a_3, \langle a_4, a_5, \langle a_6 \rangle^+ \rangle^+ \right\rangle^+$: An Ambulance goes to a Location, gets Patients on, goes to a Hospital, gets the patients off, which go to the hospital, and the hospital sends Bills to each patient. And it repeats itself.

(7) $B_7 = \langle a_1, \langle a_2 \rangle^+, \langle a_3, a_4, a_5, a_6 \rangle^+ \rangle^+$: An Ambulance goes to a Location, gets Patients on, goes to Hospitals, to get some of the patients off until all the patients off, each of which groups goes into its hospital, and the hospital sends a Bill to each patient. And it repeats itself.

(8) $B_8 = \left\langle a_1, \langle a_2 \rangle^+, \langle a_3, a_4, a_5, \langle a_6 \rangle^+ \rangle^+ \right\rangle^+$: An Ambulance goes to a Location, gets Patients on, goes to Hospitals, to get some of the patients off until all the patients off, each of which groups goes into its hospital, and the hospital sends Bills to each Patient. And it repeats itself.

(9) $B_9 = \langle a_1, \langle a_2 \rangle^+, a_3, \langle a_4, a_5, a_6 \rangle^+ | \langle a_3, a_4, a_5, a_6 \rangle^+ \rangle^+$: A repeating behavior of B_5, B_7.

(10) $B_{10} = \left\langle a_1, \langle a_2 \rangle^+, a_3, \langle a_4, a_5, \langle a_6 \rangle^+ \rangle^+ | \langle a_3, a_4, a_5, \langle a_6 \rangle^+ \rangle^+ \right\rangle^+$: A repeating behavior of B_6, B_8.

(11) $B_{11} = \langle \langle a_1, a_2 \rangle^+, a_3, \langle a_4, a_5, a_6 \rangle^+ \rangle^+$: An Ambulance goes to Locations, gets Patients on, goes to a Hospital, gets the patients off, who go to the hospital, and the hospital sends a Bill to each Patient. And it repeats itself.

(12) $B_{12} = \langle \langle a_1, a_2 \rangle^+, a_3, \langle a_4, a_5, a_6 \rangle^+ \rangle^+$: An Ambulance goes to Locations, gets Patients on, goes to Hospitals, to get some of the patients off until all the patients off, each of which groups goes into its hospital, and the hospital sends a Bill to each patient. And it repeats itself.

(13) $B_{13} = \left\langle \langle a_1, a_2 \rangle^+, a_3, \langle a_4, a_5, \langle a_6 \rangle^+ \rangle^+ \right\rangle^+$: An Ambulance goes to Locations, gets Patients on, goes to a Hospital, gets the patients off, which go to the hospital, and the hospital sends Bills to each Patient. And it repeats itself.

(14) $B_{14} = \left\langle \langle a_1, a_2 \rangle^+, \langle a_3, a_4, a_5, \langle a_6 \rangle^+ \rangle^+ \right\rangle^+$: An Ambulance goes to Locations, gets Patients on, goes to Hospitals, to get some of the patients off until all the patients off, each of which groups goes into its hospitals, and the hospital sends a Bills to each Patient. And it repeats itself.

(15) $B_{15} = \left\langle \langle a_1, a_2 \rangle^+, a_3, \langle a_4, a_5, a_6 \rangle^+ | \langle a_3, a_4, a_5, a_6 \rangle^+ \right\rangle^+$: A repeating behavior of B_{11}, B_{12}.

(16) $B_{16} = \left\langle \langle a_1, a_2 \rangle^+, a_3, \langle a_4, a_5, \langle a_6 \rangle^+ \rangle^+ | \langle a_3, a_4, a_5, \langle a_6 \rangle^+ \rangle^+ \right\rangle^+$: A repeating behavior of B_{13}, B_{14}.

(17) $B_{17} = \left\langle a_1, \langle a_2 \rangle^+ | \langle a_1, a_2 \rangle^+, a_3, \langle a_4, a_5, a_6 \rangle^+ | \langle a_3, a_4, a_5, a_6 \rangle^+ \right\rangle^+$: A repeating behavior of B_9, B_{15}.

(18) $B_{18} = \left\langle a_1, \langle a_2 \rangle^+ | \langle a_1, a_2 \rangle^+, a_3, \langle a_4, a_5, \langle a_6 \rangle^+ \rangle^+ | \langle a_3, a_4, a_5, \langle a_6 \rangle^+ \rangle^+ \right\rangle^+$: A repeating behavior of B_{10}, B_{16}.

B. HCS for B (1, n)

There are total 4 behaviors possible and are defined in regular expression as follows:

(1) $B_1 = \langle c_1, c_2, c_3 \rangle$: A customer calls an Insurance Company, then sends a Bill to the Company, and the Company pays the Bill.

(2) $B_2 = \langle c_1, \langle c_2, c_3 \rangle^+ \rangle^+$: A customer calls to an Insurance Company, then sends some Bills to the Company, and the Company pays the Bills.

(3) $B_3 = \langle \langle c_1, c_2, c_3 \rangle^+ \rangle^+$: A customer calls an Insurance Company, then sends a Bill to the Company, and the Company pays the Bill. And it repeats itself.

(4) $B_4 = \langle c_1, \langle c_2, c_3 \rangle^+ | \langle c_1, c_2, c_3 \rangle^+ \rangle^+$: A repeating behavior of B_2, B_3.

Note that regular behaviors only for EMS B(n, 1, n, n) and HCS B(1, n) are presented here due to the size of the example. However similar approach can be made for n actors.

2.3 Step 3: Abstract Behaviors

In the second step, the regular behaviors from Step 2 are abstracted with respect to a number of actors and their capacity as follows:

- Cardinality: The number of actors in behavior.
- Capacity: A number of things that an actor can handle in behavior.

Note that there are three levels of notations in order to represent the hierarchical structure of abstract behaviors. It makes mathematical composition of abstract behaviors possible for specific actors in two different domains with respect to same cardinality and capacity of the actors.

A. EMS for B (n, 1, n, n)

18 regular behaviors are abstracted in three levels as shown in Table 1. Note that the notions for each level are defined as follows:

- Level 1: A_5^3 $\begin{array}{l}\rightarrow \textit{Number of Ambulances} \\ \rightarrow \textit{Number of Patients}\end{array}$

- Level 2: $A_{(1,2,2)}^{(3)}$ $\begin{array}{l}\rightarrow \textit{Cardinality of Ambulance} \\ \rightarrow \textit{Capacity of Ambulances for patients}\end{array}$

- Level 3: $A_{\langle[1],[2,3],[4,5]\rangle}^{\langle[1],[2],[3]\rangle}$ $\begin{array}{l}\rightarrow \textit{Ambulance ID} \\ \rightarrow \textit{Patient ID}\end{array}$

B. HCS for B (1, n)

4 regular behaviors are abstracted in three levels as shown in Table 2.

Note that only abstract behaviors for EMS B(n, 1, n, n) and HCS B(1, n) are presented here due to the size of the example. However similar approach can be made for n actors.

2.4   Step 4: Abstract Behavior Lattice (ABL)

Lattice can be constructed from Step 3, based on the inclusion relations among behaviors. Formal definitions for the lattice are reported in [3]. Here we present the lattices for EMS and HSC examples.

A. EMS for B (n, 1, n, n) and HCS for B (1, n)

The Table 3 shows the inclusion relations in EMS for B(n, 1, n, n) and HCS for B(1, n), from which the abstract behavior lattices for EMS B(n, 1, n, n) and HCS B(1, n) are constructed as shown in Fig. 2.

2.5   Step 5: Composition

The last step is to make composition of two lattices for EMS and HCS. The steps of the composition are as follows:

(1) Firstly, the common actors between two abstract behavior lattices have to be selected. For the example, Patient from EMS is defined to be a common actor with Customer from HCS.

(2) Secondly, cardinality of the composition has to be selected for the common actors. For the example, there are two cases: one for the single cardinality and the other for the plural cardinality.

A. Composition for $EMS(L, A, H, P) \otimes_{EMS(P)=HCS(C)\&|P|=|C|=1} HCS(C, I)$

This is the first case of the composition for EMS and HCS with respect to Patient of Cardinality 1: $EMS(L, A, H, P) \otimes_{EMS(P)=HCS(C)\&|P|=|C|=1} HCS(C, I)$. $EMS(P) = HCS(C)$ implies that Patient from EMS is defined to be a common actor with Customer from HCS, and $|P| = |C| = 1$ implies that their cardinality is singular. Figure 3a. shows the possible composition for EMS and HCS, and Fig. 3b. shows that result of the composition: $\left(EMS(L, A, H, P) \otimes_{EMS(P)=HCS(C)\&|P|=|C|=1} HCS(C, I)\right)(L, A, H, P =$

Table 1. Abstract behaviors for EMS B(n, 1, n, n)

Level 1

$B_1\left(L_1^1, A_1^1, H_1^1, P_1^1\right)$	$B_2\left(L_1^1, A_1^1, H_1^1, P_m^1\right)$	$B_3\left(L_x^i, A_1^1, H_z^k, P_n^n\right)$	$B_4\left(L_x^i, A_1^1, H_z^k, P_m^n\right)$
$B_5\left(L_x^1, A_y^1, H_z^1, P_n^n\right)$	$B_6\left(L_x^1, A_y^1, H_z^1, P_m^n\right)$	$B_7\left(L_x^1, A_y^1, H_k^k, P_n^n\right)$	$B_8\left(L_x^1, A_y^1, H_k^k, P_m^n\right)$
$B_9\left(L_x^1, A_y^1, H_z^k, P_n^n\right)$	$B_{10}\left(L_x^1, A_y^1, H_z^k, P_m^n\right)$	$B_{11}\left(L_i^i, A_{ny}^1, H_z^1, P_n^n\right)$	$B_{12}\left(L_i^i, A_y^1, H_k^k, P_n^n\right)$
$B_{13}\left(L_i^i, A_y^1, H_z^1, P_m^n\right)$	$B_{14}\left(L_i^i, A_y^1, H_k^k, P_n^n\right)$	$B_{15}\left(L_i^i, A_y^1, H_z^k, P_n^n\right)$	$B_{16}\left(L_i^i, A_y^1, H_z^k, P_m^n\right)$
$B_{17}\left(L_x^i, A_y^1, H_z^k, P_n^n\right)$	$B_{18}\left(L_x^i, A_y^1, H_z^k, P_m^n\right)$		

Level 2

$B_1\left(L_{\langle1\rangle}^{\langle1\rangle}, A_{\langle1\rangle}^{\langle1\rangle}, H_{\langle1\rangle}^{\langle1\rangle}, P_{\langle1\rangle}^{\langle1\rangle}\right)$	$B_2\left(L_{\langle1\rangle}^{\langle1\rangle}, A_{\langle1\rangle}^{\langle1\rangle}, H_{\langle1\rangle}^{\langle1\rangle}, P_{\langle m\rangle}^{\langle1\rangle}\right)$
$B_3\left(L_{\langle x_1\dots x_i\rangle}^{\langle i\rangle}, A_{\langle1\rangle}^{\langle1\rangle}, H_{\langle z_1\dots z_k\rangle}^{\langle k\rangle}, P_{\langle1_1\dots1_n\rangle}^{\langle n\rangle}\right)$	$B_4\left(L_{\langle x_1\dots x_i\rangle}^{\langle i\rangle}, A_{\langle1\rangle}^{\langle1\rangle}, H_{\langle z_1\dots z_k\rangle}^{\langle k\rangle}, P_{\langle m_1\dots m_n\rangle}^{\langle n\rangle}\right)$
$B_5\left(L_{\langle x\rangle}^{\langle1\rangle}, A_{\langle y\rangle}^{\langle1\rangle}, H_{\langle z\rangle}^{\langle1\rangle}, P_{\langle1_1\dots1_n\rangle}^{\langle n\rangle}\right)$	$B_6\left(L_{\langle x\rangle}^{\langle1\rangle}, A_{\langle y\rangle}^{\langle1\rangle}, H_{\langle z\rangle}^{\langle1\rangle}, P_{\langle m_1\dots m_n\rangle}^{\langle n\rangle}\right)$
$B_7\left(L_{\langle x\rangle}^{\langle1\rangle}, A_{\langle y\rangle}^{\langle1\rangle}, H_{\langle1_1\dots1_k\rangle}^{\langle k\rangle}, P_{\langle1_1\dots1_n\rangle}^{\langle n\rangle}\right)$	$B_8\left(L_{\langle x\rangle}^{\langle1\rangle}, A_{\langle y\rangle}^{\langle1\rangle}, H_{\langle1_1\dots1_k\rangle}^{\langle k\rangle}, P_{\langle m_1\dots m_n\rangle}^{\langle n\rangle}\right)$
$B_9\left(L_{\langle x\rangle}^{\langle1\rangle}, A_{\langle y\rangle}^{\langle1\rangle}, H_{\langle z_1\dots z_k\rangle}^{\langle k\rangle}, P_{\langle1_1\dots1_n\rangle}^{\langle n\rangle}\right)$	$B_{10}\left(L_{\langle x\rangle}^{\langle1\rangle}, A_{\langle y\rangle}^{\langle1\rangle}, H_{\langle z_1\dots z_k\rangle}^{\langle k\rangle}, P_{\langle m_1\dots m_n\rangle}^{\langle n\rangle}\right)$
$B_{11}\left(L_{\langle1_1\dots1_i\rangle}^{\langle i\rangle}, A_{\langle y\rangle}^{\langle1\rangle}, H_{\langle z\rangle}^{\langle1\rangle}, P_{\langle1_1\dots1_n\rangle}^{\langle n\rangle}\right)$	$B_{12}\left(L_{\langle1_1\dots1_i\rangle}^{\langle i\rangle}, A_{\langle y\rangle}^{\langle1\rangle}, H_{\langle1_1\dots1_k\rangle}^{\langle k\rangle}, P_{\langle1_1\dots1_n\rangle}^{\langle n\rangle}\right)$
$B_{13}\left(L_{\langle1_1\dots1_i\rangle}^{\langle i\rangle}, A_{\langle y\rangle}^{\langle1\rangle}, H_{\langle z\rangle}^{\langle1\rangle}, P_{\langle m_1\dots m_n\rangle}^{\langle n\rangle}\right)$	$B_{14}\left(L_{\langle1_1\dots1_i\rangle}^{\langle i\rangle}, A_{\langle y\rangle}^{\langle1\rangle}, H_{\langle1_1\dots1_k\rangle}^{\langle k\rangle}, P_{\langle m_1\dots m_n\rangle}^{\langle n\rangle}\right)$
$B_{15}\left(L_{\langle1_1\dots1_i\rangle}^{\langle i\rangle}, A_{\langle y\rangle}^{\langle1\rangle}, H_{\langle z_1\dots z_k\rangle}^{\langle k\rangle}, P_{\langle1_1\dots1_n\rangle}^{\langle n\rangle}\right)$	$B_{16}\left(L_{\langle1_1\dots1_i\rangle}^{\langle i\rangle}, A_{\langle y\rangle}^{\langle1\rangle}, H_{\langle z_1\dots z_k\rangle}^{\langle k\rangle}, P_{\langle m_1\dots m_n\rangle}^{\langle n\rangle}\right)$
$B_{17}\left(L_{\langle x_1\dots x_i\rangle}^{\langle i\rangle}, A_{\langle y\rangle}^{\langle1\rangle}, H_{\langle z_1\dots z_k\rangle}^{\langle k\rangle}, P_{\langle1_1\dots1_n\rangle}^{\langle n\rangle}\right)$	$B_{18}\left(L_{\langle x_1\dots x_i\rangle}^{\langle i\rangle}, A_{\langle y\rangle}^{\langle1\rangle}, H_{\langle z_1\dots z_k\rangle}^{\langle k\rangle}, P_{\langle m_1\dots m_n\rangle}^{\langle n\rangle}\right)$

Level 3

$B_1\left(L_{\langle[p]\rangle}^{\langle[l]\rangle}, A_{\langle[p]\rangle}^{\langle[a]\rangle}, H_{\langle[p]\rangle}^{\langle[h]\rangle}, P_{\langle[b]\rangle}^{\langle[p]\rangle}\right)$

$B_2\left(L_{\langle[p]\rangle}^{\langle[l]\rangle}, A_{\langle[p]\rangle}^{\langle[a]\rangle}, H_{\langle[p]\rangle}^{\langle[h]\rangle}, P_{\langle[b_1\dots b_m]\rangle}^{\langle[p]\rangle}\right)$

$B_3\left(L_{\langle[p_1\dots p_{x_1}]_1\dots[p_1\dots p_{x_i}]_i\rangle}^{\langle[l]_1\dots[l]_i\rangle}, A_{\langle[p]\rangle}^{\langle[a]\rangle}, H_{\langle[p_1\dots p_{z_1}]_1\dots[p_1\dots p_{z_k}]_k\rangle}^{\langle[h]_1\dots[h]_k\rangle}, P_{\langle[b]_1\dots[b]_n\rangle}^{\langle[p]_1\dots[p]_n\rangle}\right)$

$B_4\left(L_{\langle[p_1\dots p_{x_1}]_1\dots[p_1\dots p_{x_i}]_i\rangle}^{\langle[l]_1\dots[l]_i\rangle}, A_{\langle[p]\rangle}^{\langle[a]\rangle}, H_{\langle[p_1\dots p_{z_1}]_1\dots[p_1\dots p_{z_k}]_k\rangle}^{\langle[h]_1\dots[h]_k\rangle}, P_{\langle[b_1\dots b_{m_1}]_1\dots[b_1\dots b_{m_n}]_n\rangle}^{\langle[p]_1\dots[p]_n\rangle}\right)$

$B_5\left(L_{\langle[p_1\dots p_x]\rangle}^{\langle[l]\rangle}, A_{\langle[p_1\dots p_y]\rangle}^{\langle[a]\rangle}, H_{\langle[p_1\dots p_z]\rangle}^{\langle[h]\rangle}, P_{\langle[b]_1\dots[b]_n\rangle}^{\langle[p]_1\dots[p]_n\rangle}\right)$

$B_6\left(L_{\langle[p_1\dots p_x]\rangle}^{\langle[l]\rangle}, A_{\langle[p_1\dots p_y]\rangle}^{\langle[a]\rangle}, H_{\langle[p_1\dots p_z]\rangle}^{\langle[h]\rangle}, P_{\langle[b_1\dots b_{m_1}]_1\dots[b_1\dots b_{m_n}]_n\rangle}^{\langle[p]_1\dots[p]_n\rangle}\right)$

$B_7\left(L_{\langle[p_1\dots p_x]\rangle}^{\langle[l]\rangle}, A_{\langle[p_1\dots p_y]\rangle}^{\langle[a]\rangle}, H_{\langle[p_1\dots p]_k\rangle}^{\langle[h]_1\dots[h]_k\rangle}, P_{\langle[b]_1\dots[b]_n\rangle}^{\langle[p]_1\dots[p]_n\rangle}\right)$

$B_8\left(L_{\langle[p_1\dots p_x]\rangle}^{\langle[l]\rangle}, A_{\langle[p_1\dots p_y]\rangle}^{\langle[a]\rangle}, H_{\langle[p_1\dots p]_k\rangle}^{\langle[h]_1\dots[h]_k\rangle}, P_{\langle[b_1\dots b_{m_1}]_1\dots[b_1\dots b_{m_n}]_n\rangle}^{\langle[p]_1\dots[p]_n\rangle}\right)$

$B_9\left(L_{\langle[p_1\dots p_x]\rangle}^{\langle[l]\rangle}, A_{\langle[p_1\dots p_y]\rangle}^{\langle[a]\rangle}, H_{\langle[p_1\dots p_{z_1}]_1\dots[p_1\dots p_{z_k}]_k\rangle}^{\langle[h]_1\dots[H]_k\rangle}, P_{\langle[b]_1\dots[b]_n\rangle}^{\langle[p]_1\dots[p]_n\rangle}\right)$

(continued)

Table 1. (*continued*)

$$B_{10}\left(L_{\langle[p_1\dots p_x]\rangle}^{\langle[l]\rangle}, A_{\langle[p_1\dots p_y]\rangle}^{\langle[a]\rangle}, H_{\langle[p_1\dots p_{z_1}]_1\dots[p_1\dots p_{z_k}]_k\rangle}^{\langle[h]_1\dots[h]_k\rangle}, P_{\langle[b_1\dots b_{m_1}]_1\dots[b_1\dots b_{mn}]_n\rangle}^{\langle[p]_1\dots[p]_n\rangle}\right)$$

$$B_{11}\left(L_{\langle[p]_1\dots[p]_n\rangle}^{\langle[l]_1\dots[l]_i\rangle}, A_{\langle[p_1\dots p_y]\rangle}^{\langle[a]\rangle}, H_{\langle[p_1\dots p_z]\rangle}^{\langle[h]\rangle}, P_{\langle[b]_1\dots[b]_n\rangle}^{\langle[p]_1\dots[p]_n\rangle}\right)$$

$$B_{12}\left(L_{\langle[p]_1\dots[p]_n\rangle}^{\langle[l]_1\dots[l]_i\rangle}, A_{\langle[p_1\dots p_y]\rangle}^{\langle[a]\rangle}, H_{\langle[p]_1\dots[p]_k\rangle}^{\langle[h]_1\dots[h]_k\rangle}, P_{\langle[b]_1\dots[b]_n\rangle}^{\langle[p]_1\dots[p]_n\rangle}\right)$$

$$B_{13}\left(L_{\langle[p]_1\dots[p]_n\rangle}^{\langle[l]_1\dots[l]_i\rangle}, A_{\langle[p_1\dots p_y]\rangle}^{\langle[a]\rangle}, H_{\langle[p_1\dots p_z]\rangle}^{\langle[h]\rangle}, P_{\langle[b_1\dots b_{m_1}]_1\dots[b_1\dots b_{mn}]_n\rangle}^{\langle[p]_1\dots[p]_n\rangle}\right)$$

$$B_{14}\left(L_{\langle[p]_1\dots[p]_n\rangle}^{\langle[l]_1\dots[l]_i\rangle}, A_{\langle[p_1\dots p_y]\rangle}^{\langle[a]\rangle}, H_{\langle[p]_1\dots[p]_k\rangle}^{\langle[h]_1\dots[h]_k\rangle}, P_{\langle[b_1\dots b_{m_1}]_1\dots[b_1\dots b_{mn}]_n\rangle}^{\langle[p]_1\dots[p]_n\rangle}\right)$$

$$B_{15}\left(L_{\langle[p]_1\dots[p]_n\rangle}^{\langle[l]_1\dots[l]_i\rangle}, A_{\langle[p_1\dots p_y]\rangle}^{\langle[a]\rangle}, H_{\langle[p_1\dots p_{z_1}]_1\dots[p_1\dots p_{z_k}]_k\rangle}^{\langle[h]_1\dots[h]_k\rangle}, P_{\langle[b]_1\dots[b]_n\rangle}^{\langle[p]_1\dots[p]_n\rangle}\right)$$

$$B_{16}\left(L_{\langle[p]_1\dots[p]_n\rangle}^{\langle[l]_1\dots[l]_i\rangle}, A_{\langle[p_1\dots p_y]\rangle}^{\langle[a]\rangle}, H_{\langle[p_1\dots p_{z_1}]_1\dots[p_1\dots p_{z_k}]_k\rangle}^{\langle[h]_1\dots[h]_k\rangle}, P_{\langle[b_1\dots b_{m_1}]_1\dots[b_1\dots b_{mn}]_n\rangle}^{\langle[p]_1\dots[p]_n\rangle}\right)$$

$$B_{17}\left(L_{\langle[p_1\dots p_{x_1}]_1\dots[p_1\dots p_{x_i}]_i\rangle}^{\langle[l]_1\dots[l]_i\rangle}, A_{\langle[p_1\dots p_y]\rangle}^{\langle[a]\rangle}, H_{\langle[p_1\dots p_{z_1}]_1\dots[p_1\dots p_{z_k}]_k\rangle}^{\langle[h]_1\dots[h]_k\rangle}, P_{\langle[b]_1\dots[b]_n\rangle}^{\langle[p]_1\dots[p]_n\rangle}\right)$$

$$B_{18}\left(L_{\langle[p_1\dots p_{x_1}]_1\dots[p_1\dots p_{x_i}]_i\rangle}^{\langle[l]_1\dots[l]_i\rangle}, A_{\langle[p_1\dots p_y]\rangle}^{\langle[a]\rangle}, H_{\langle[p_1\dots p_{z_1}]_1\dots[p_1\dots p_{z_k}]_k\rangle}^{\langle[h]_1\dots[h]_k\rangle}, P_{\langle[b_1\dots b_{m_1}]_1\dots[b_1\dots b_{mn}]_n\rangle}^{\langle[p]_1\dots[p]_n\rangle}\right)$$

$C, I)_{|P|=|C|=1} = B(n, n, n, 1, n)$. There are total 4 possible collective behaviors, which is the half of the total composition with the same cardinality, that is, 8 behaviors, 2 from EMS by 4 from HCS, and is 1/18 of the total composition with the different cardinality, that is 72 behaviors, 18 from EMS by 4 from HCS.

B. Composition for $EMS(L, A, H, P) \otimes_{EMS(P)=HCS(C) \& |P|=|C|=n} HCS(C, I)$

This case has been omitted intentionally in the paper due to the complexity of the example.

Table 2. Abstract behaviors for HCS B (1, n)

Level 1	Level 2	Level 3
$B_1\left(C_1^1, I_y^1\right)$	$B_1\left(C_{\langle1\rangle}^1, I_{\langle1\rangle}^1\right)$	$B_1\left(C_{\langle[1]\rangle}^{\langle[1]\rangle}, I_{\langle[1]\rangle}^{\langle[1]\rangle}\right)$
$B_2\left(C_x^1, I_y^1\right)$	$B_2\left(C_{\langle x\rangle}^1, I_{\langle y\rangle}^1\right)$	$B_2\left(C_{\langle[1]\dots[x]\rangle}^{\langle[1]\rangle}, I_{\langle[1]\dots[y]\rangle}^{\langle[1]\rangle}\right)$
$B_3\left(C_x^1, I_j^j\right)$	$B_3\left(C_{\langle x\rangle}^1, I_{\langle1_1\dots1_j\rangle}^j\right)$	$B_3\left(C_{\langle[1]\dots[x]\rangle}^{\langle[1]\rangle}, I_{\langle[1]\dots[j]\rangle}^{\langle[1]\dots[j]\rangle}\right)$
$B_4\left(C_x^1, I_y^j\right)$	$B_4\left(C_{\langle x\rangle}^1, I_{\langle y_1\dots y_j\rangle}^j\right)$	$B_4\left(C_{\langle[1]\dots[x]\rangle}^{\langle[1]\rangle}, I_{\langle[y_1]\dots[y_j]\rangle}^{\langle[1]\dots[j]\rangle}\right)$

Table 3. Inclusion relations for EMS B(n. 1, n, n) and HCS B(1,n)

EMS							HCS
$B_1 \sqsubseteq B_2$	$B_1 \sqsubseteq B_{13}$	$B_2 \sqsubseteq B_{14}$	$B_5 \sqsubseteq B_9$	$B_9 \sqsubseteq B_{10}$	$B_{12} \sqsubseteq B_{16}$	$B_{15} \sqsubseteq B_{17}$	$B_1 \sqsubseteq B_2$
$B_1 \sqsubseteq B_3$	$B_2 \sqsubseteq B_3$	$B_3 \sqsubseteq B_4$	$B_6 \sqsubseteq B_{10}$	$B_9 \sqsubseteq B_{17}$	$B_{13} \sqsubseteq B_{14}$	$B_{16} \sqsubseteq B_{18}$	$B_1 \sqsubseteq B_3$
$B_1 \sqsubseteq B_5$	$B_2 \sqsubseteq B_4$	$B_3 \sqsubseteq B_{17}$	$B_7 \sqsubseteq B_8$	$B_{10} \sqsubseteq B_{18}$	$B_{13} \sqsubseteq B_{15}$	$B_{17} \sqsubseteq B_{18}$	$B_1 \sqsubseteq B_4$
$B_1 \sqsubseteq B_7$	$B_2 \sqsubseteq B_8$	$B_4 \sqsubseteq B_{18}$	$B_7 \sqsubseteq B_9$	$B_{11} \sqsubseteq B_{12}$	$B_{14} \sqsubseteq B_{16}$		$B_2 \sqsubseteq B_4$
$B_1 \sqsubseteq B_{11}$	$B_2 \sqsubseteq B_{12}$	$B_5 \sqsubseteq B_6$	$B_8 \sqsubseteq B_{10}$	$B_{11} \sqsubseteq B_{15}$	$B_{15} \sqsubseteq B_{16}$		$B_3 \sqsubseteq B_4$

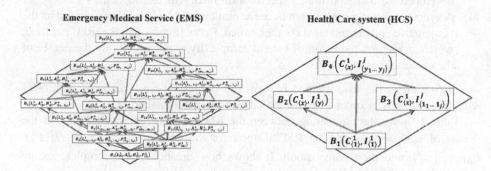

Fig. 2. Abstract behavior Lattice for EMS B(n, 1, n, n) and HCS B(1, n)

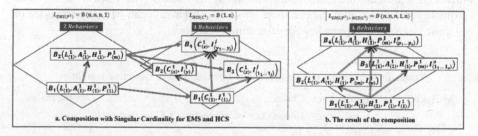

Fig. 3. Composition with singular cardinality for EMS and HCS

3   Analysis and Comparison

This paper presented a new approach to reduce system complexity based on the new notion of Abstract Behaviors Lattice (ABL). It guarantees that the complexity is reduced exponentially by the definition of the *abstract behavior* lattice. Further, this method represents composition of two systems with common actors in the minimum states.

There are a number of approaches to reduce system complexities. The best known approaches are as follows:

(1) A compositional analysis of finite state systems to deal with state explosion due to process composition: They tried to reduce base on synchronous and asynchronous execution and showed the method by process algebra [5, 6].
(2) A technique to cluster states into equivalent classes which uses a graphical representation into text form [7]: It uses Communicating Real-Time State Machines (CRSMs) that works on automatic verification of finite state real-time systems.
(3) A technique to reduce possible infinite time space into finite time space which is developed for a compositional specification theory for components [8].
(4) A syntactic sugar approach to reduce the number of synchronizations based on the conjunctive and complement choices, called, Reduction-by-Choice (RbC) method, which guarantees reduction of system complexity exponentially by the degree of the choice dependences [1].

Among these, the RbC method is known to be best: it was compared with other reduction methods in order to prove the efficiency of the method.

Table 4 shows the comparing between the RbC and our approach. It shows the results of comparison with the RbC method for the case of $B_{EMS}\left(L_4^2, A_4^2, H_4^2, P_4^4\right)$, $B_{HCS}\left(C_4^4, I_4^1\right)$, and their composition. It shows how drastically the complexities are reduced by the ABL method.

Compared to the other approaches, the ABL approach reduces the states with respect to the types of behaviors, that is, a sequence of interactions among actors. Further, it represents the composition of two system states with respect to the same cardinality and capacity of the common actors. The approach can be considered one of the most innovative approaches for state minimization.

Table 4. Comparing choice method and our method in the same condition

Methods	Original states	Choice		Our methods
		Conjunction	Conjunction + Complement	
EMS	746496	36	6	4
HCS	–	–	–	1
EMS ⊗ HCS	–	≥ 36	≥ 6	4

4 Prism

In order to demonstrate the feasibility of the approach in this paper, a tool, called PRISM, has been developed on ADOxx Meta-Modeling Platform [4]. As shown in Fig. 4, it consists of the basic models for each step of the approach in the paper: Class Diagram Model, Regular Behavior Model, Abstract Behavior Model, Abstract Behavior Lattice Model, and Composition Model. Further there are a number of engines to support functionality of the model, such as, parsers, transformers, extractors, abstractors, mappers, etc.

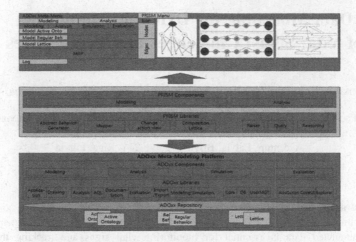

Fig. 4. The views and architecture of PRISM on ADOxx

5   Conclusion and Future Research

This paper presented a new method for knowledge engineering and composition to model collective behaviours of systems, based on Behavior Otology. The efficiency of the method was shown with numbers for reduction of the system state. Finally, the PRISM tool was described at the end. The future research includes developing meta-modeling the method and its instantiation to target domains, developing an open model tool for PRISM on ADOxx, providing sound and complete proof for the method.

References

1. Choi, W., Choe, Y., Lee, M.: A reduction method for process and system complexity with conjunctive and complement choices in a process algebra. In: 39th IEEE COMPSAC/MVDM, July 2015
2. Woo, S., On, J., Lee, M.: An abstraction method for mobility, and interaction in process algebra using behavior ontology. In: 37th IEEE COMPSAC, July 2013
3. Choe, Y., Lee, M.: A Lattice model to verify behavioral equivalence. In: UKSim-AMSS 8th European Modelling Symposium, October 2014
4. Fill, H., Karagiannis, D.: On the conceptualisation of modeling methods using the ADOxx meta modeling platform. Enterp. Model. Inf. Syst. Architect. **8**(1), 4–25 (2013)
5. Clarke, E., Emerson, A., Sifakis, J.: Model checking: algorithmic verification and debugging. Commun. ACM **52**(11), 74–84 (2009)
6. Yeh, W., Young, M.: Compositional reachability analysis using process algebra. In: Proceedings of Conference on Testing, Analysis and Verification, pp. 49–59, August 1992
7. Chen, T., Chilton, C., Jonsson, B., Kwiatkowska, M.: A compositional specification theory for component behaviours. In: Seidl, H. (ed.) ESOP 2012. LNCS, vol. 7211, pp. 148–168. Springer, Heidelberg (2012). https://doi.org/10.1007/978-3-642-28869-2_8
8. Raju, S.: An automatic verification technique for communicating real-time state machines. Technical report 93-07-08, Department of Computer science and Engineering, University of Washington, April 1993

Data Modelling for Dynamic Monitoring of Vital Signs: Challenges and Perspectives

Natalija Kozmina[1(⊠)], Emil Syundyukov[1], and Aleksejs Kozmins[2]

[1] Faculty of Computing, University of Latvia, Raina blvd. 19, Riga, Latvia
natalija.kozmina@lu.lv, e.syundyukov@gmail.com
[2] Accenture Latvia, Brivibas gatve 214, Riga, Latvia
aleksejs.kozmins@accenture.com

Abstract. The use-case described in this paper covers data acquisition and real-time analysis of the gathered medical data from wearable sensor system. Accumulated data is essential for monitoring vital signs and tracking the dynamics of the treatment process of disabled patients or patients undergoing the recovery after traumatic knee joint injury (e.g. post-operative rehabilitation). The main goal of employing the wearable sensor system is to conduct rehabilitation process more effectively and increase the rate of successful rehabilitation. The results of data analysis of patient's vital signs and feedback allow a physiotherapist to adjust the rehabilitation scenario on the fly. In this paper, we focus on the methodology for data modelling with a purpose to design a computer-aided rehabilitation system that would support agility of changing information requirements by being flexible and augmentable.

Keywords: Knee joint · Rehabilitation · Real-time monitoring · Data modelling

1 Introduction

One of the most common cases of injuries that occur in people of all ages is an injury of the knee joint. As reported in [1], in the year of 2010 there were roughly 10.4 million patients' visits to doctors' offices because of common knee injuries such as fractures, dislocations, sprains, and ligament tears. In most of the cases, a surgical assistance is required for healing process. Then a rehabilitation routine follows that is aimed to minimize swelling and return the range of movement as well as to strengthen leg muscles by taking into account limitations set by a physiotherapist (e.g. flexion limitation to 90°, length of the rehabilitation session, etc.).

Rehabilitation is one of the most important parts of the treatment process. Laboratory-based equipment could be used to collect data on angular motion, and though this kind of research advances valuable information, the results remain valid only in conditions where, no anticipation or reaction to the real world environment is required. According to [2], a body sensor network consisting of wireless sensors attached to a patient provides a promising method to collect clinically relevant information about knee function in everyday situations. In [3], it is recommended that healthcare specialists in the hospital would supervise the rehabilitation process,

© Springer International Publishing AG 2017
S. de Cesare and U. Frank (Eds.): ER 2017 Workshops, LNCS 10651, pp. 16–25, 2017.
https://doi.org/10.1007/978-3-319-70625-2_2

however, patients should undergo necessary procedures on their own taking doctor's prescriptions into consideration. Continuous real-time monitoring can be achieved by using wearable devices and wireless sensor networks. Meanwhile, long-term monitoring of the physiological data could lead to significant improvement in the diagnosis and treatment of the diseases as stated in [2].

Real-time data analysis is critical to perform diagnostics and trace the dynamics in changes of the patient's state of health. Typically, data to be ingested is acquired in real time from a set of wireless sensor networks consisting of embedded devices with inertial measurement unit (IMU) and other sensor units. Our developed system aims to fill in the gap and provide an approach for patient post-operative rehabilitation monitoring. The system consists of a combination of embedded devices, sensor nodes for data acquisition, and a mobile application for data storage (NoSQL database) and analysis. The mobile application provides data to a web-based framework for data visualisation, analytics, and Cloud-based storage. In this paper, we concentrate on the aspect of data modelling suitable for performing both real-time and a more complex historical analysis of patients' data and discovered challenges that persist.

The remaining part of the paper is organized as follows. Section 2 gives a summary of related work both in the area of knee joint rehabilitation monitoring and data modelling for such systems, Sect. 3 presents a short overview of our framework, Sect. 4 describes our envisioned approach for data model re-design, and Sect. 5 finalizes the paper with conclusions and future work in different directions.

2 Related Work

To explore the existing wearable sensor systems used for knee joint rehabilitation monitoring and find possible gaps in this area, we accomplished a small-scale literature study. We performed the search via Google Scholar and IEEE Sensors Journal databases, and limited the timeframe to the last ten years to get the most recent papers (e.g. data from proposed systems is compared to new technological product like Microsoft Kinect or XSens Technologies systems). We used the following search string in Google Scholar: *"biofeedback" AND "rehabilitation" AND "wearable" AND "sensor" AND "knee joint" AND (year >= 2006 AND year <= 2016) NOT patents.*

In IEEE Sensors Journal we set the search restrictions to: Keywords = knee motion; Journal = sensors. We reviewed to most relevant 100 papers out of 483 found in Google Scholar and all 8 retrieved from IEEE Sensors. A part of the papers were eliminated because the search keywords were used in a different context (e.g. non-computer science field, systems developed for other applications, literature reviews, surveys). Snowballing applied for the paper with the highest number of citations [4] resulted in 15 papers for further analysis.

We were interested in investigating the following research questions. *RQ1:* What type of biofeedback is provided to the patient during rehabilitation session? *RQ2:* What kind of sensors is used for knee joint rehabilitation monitoring? *RQ3:* What type of communication/feedback is returned to healthcare specialists? Regarding *RQ1*, Foody et al. [5] described a biofeedback mechanism and proposed a video game that is capable of providing audible and visual feedback/instructions to a user. Matsubara et al.

[6] analysed the impact of biofeedback given to patients during rehabilitation sessions. To answer *RQ2*, mainly IMU sensors for data acquisition are employed for dynamics monitoring. Five studies included only accelerometers and gyroscopes [7–11]. For example, Seel et al. [11] stated that they avoided using magnetometers in their system, so that it would not rely on a homogeneous magnetic field. In [12], gyroscopes were avoided, and data acquired exclusively from accelerometers and magnetometers was combined. Yamada et al. [7] presented a system that had conductive textile sensors for data acquisition. They proposed a class of wearable and stretchable devices fabricated from thin films of aligned single-walled carbon nanotubes. Considering *RQ3* on communication/feedback, Daponte et al. [12] claimed that clinical staff is able to remotely monitor the movements of the subject via 3D reconstruction. Yurtman et al. [13] stated that the feedback could also be in the form of the notification alerts to inform physicians/therapists only when needed.

Though other research groups have conducted a number of studies, we noticed that little attention is paid to the data modelling as such. It yielded another research question, *RQ4:* Is there a data modelling approach to enhance both real-time and historical data analysis? While looking for relevant papers on data model design for use-cases similar to ours, we performed a search in Google Scholar by the following string: *("data model") AND ("post-traumatic" OR "post-operative") AND - "post-traumatic stress disorder" AND rehab**. However, out of 72 sources no relevant work regarding data modelling for rehabilitation procedures was found. The typical approach for storing data is a traditional relational model. In [14] an elaborate description of the relational database to store human motion data is presented. Nevertheless, either of the sources lacks discussion about handling the data model if it evolves over time.

In our paper, we put an emphasis on selection of the approach for data storage guided by a set of restrictions of the system. The restrictions include the need for both real-time and historical data analysis, frequently changing information requirements and, as a result, new data objects and structures, and so on.

3   Framework Architecture for Knee Joint Dynamics Monitoring and Its Further Extensions

The initial version of the framework for knee joint dynamics monitoring is presented in [15] and depicted in Fig. 1. A physiotherapist is able to view real-time and historical analysis data of a patient's knee joint measurements, and perform an assessment of a patient's state of health and monitor dynamics of convalescence. The rehabilitation program consists of the number of stages during which a patient is prescribed to fulfil a set of exercises on a regular basis either under supervision of the physiotherapist or at home on his/her own in unattended mode. The developed framework prototype consists of wearable devices and mobile software for health data acquisition. The aim is to ensure that a physiotherapist is able to monitor patient knee joint movement angle in real time during rehabilitation procedures and adjust the rehabilitation program if necessary. A mobile application was developed to make calculations, analyse collected data from sensor nodes, store and visualize it, and provide feedback. Application functionality includes communication with a patient via real-time feedback. Alerting about the dangerous situations such as exceeding doctor prescriptions on time can help decrease the chance of repeated injuries. Notification system is implemented based on health specialist/patient communication style aiming to create a similar feeling of safety during in home rehabilitation when a patient is under direct supervision of the physiotherapist. A patient is notified automatically when exceeding the flexion limit threshold, violating the rule of exercise execution, reaching the end of rehabilitation exercise session, or the rehabilitation exercise session is over.

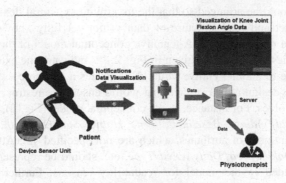

Fig. 1. Structure of proposed framework architecture for knee joint dynamics monitoring.

Another task is the quality check of the real-time data streams and employed equipment; the data received from each sensor should be verified and validated. If the produced data is incorrect (e.g. duplicated/bad/absent data), then the malfunction of a particular sensor is detected, and this sensor needs to be serviced or replaced.

Currently, the framework provides a real-time monitoring of flexion angle during two types of exercises: squats and patient's knee flexion. Wearable sensor system consists of a circuit board with MSP430 microcontroller for data sampling with 50 Hz rate and 4 sensor nodes that include sensors (3-axial accelerometers and magnetometers). For knee joint flexion/extension angle calculation a network consisting of four 3-axial accelerometers mounted on the knee is used. Precision of the developed prototype is $\pm 0.79°$ in comparison to industrial goniometer (precision: ± 0.1).

It is planned to broaden the spectrum of recovery exercises (e.g. short-arc exercise, straight leg raise as in [10]), extend the number of sensor units required for taking measurements (e.g. pulse, blood pressure), gather more patient-related data (e.g. feedback from the patient on how he/she feels during/after exercise sessions), perform statistical analysis of grouped patients' data (e.g. by cohorts of patients determined by their characteristics and habits, state of health features, past diseases). All of this leads to larger volumes of data, more sophisticated analytical tasks that may be set as the framework evolves, and implies a need to re-design a data model, altogether serving as a motivation for this study.

4 An Approach to Data Model Re-design and Its Challenges

The data has to be accumulated in the special data storage and pre-processed during ingestion process in a form suitable to perform further real-time and historical analysis of patients' data. In this section, we discuss issues and requirements for the data storage suitable for our use-case that covers the task of organizing data on patients and rehabilitation programs.

4.1 Structural Flexibility and Automatisation Capabilities

Collected data should be arranged so that the maximum structural flexibility would be feasible. The exact and final structure of data is not yet fully known, and it is not considered to be a reachable goal. A tentative conceptual model of the framework for knee joint dynamics monitoring is given in Fig. 2; '*' indicates the extensibility of the entity or reference to another entity, while '^' indicates a sub-process.

Our goal is to design a data storage that would ensure the restructuring of data as the framework evolves. For instance, in the conceptual model (Fig. 2), such entities as *Exercise*, *PhysicalFinding*, *BaselinePattern*, *Habit*, *StateOfHealth*, etc. may be extended with a number of attributes, which are not specified yet. Attributes such as *ExerciseType*, *SmokingFreq*, *DailyActivityType*, etc. should be represented as separate entities not included in the model for readability reasons. Furthermore, a similar principle applies to the data representation of the rehabilitation program in general, which is defined as a complex hierarchical structure. For example, *GeneralEvaluation*,

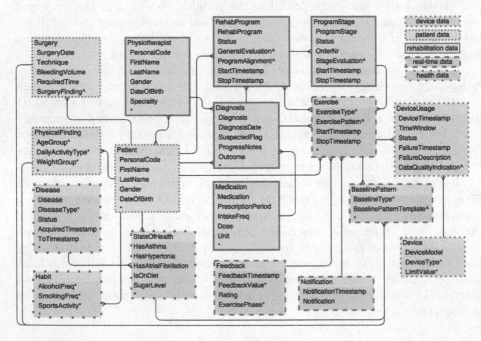

Fig. 2. Conceptual model of the framework for knee joint dynamics monitoring.

ProgramAlignment, *BaselinePatternTemplate*, *DataQualityIndication* refer to a set of entities that describe separate sub-processes of the rehabilitation program (e.g. pattern matching between patient's exercise pattern and a baseline acquired by machine learning algorithms) and are not reflected in the model.

Data structures should be used to select an appropriate rehabilitation program automatically. In fact, a mapping has to be built which takes data on *Patient*, *Diagnosis*, *Surgery*, *Disease*, *Medication*, *Habit*, *PhysicalFinding*, and *StateOfHealth* that serves as an input to suggest an appropriate rehabilitation program for each patient. Taking into consideration that the final version of the data model does not exist, we formulate a set of overall requirements that have to be met in the course of the physical implementation of the data model. It should ensure maximum simplicity of maintaining: (i) changes in the structure of all entities of the data model, (ii) changes of inter-component relations, and (iii) history of all changes made.

4.2 Complexity of the Data Model Re-design

A relational data model would ensure the implementation of the above-mentioned requirements, but still it is not adaptable to support the frequent changes: it requires significant time and effort of software developers for re-designing data structures and adjusting analytics. As adaptation to changes should be carried out unpredictably often, this approach will not give proper effectiveness and is unlikely to be justified. A "check-list" for the data model could be as follows: it should combine transactional, historical tracking/data warehousing and analytical processing features, being effort-/time-saving in development and further support.

According to the results of recent studies [16, 17], other typical data models (e.g. the ones used in data warehousing, *DW*) are not efficient enough for providing capabilities of the real-time data acquisition, whereas Data Vault (DV) model suggested by Dan Linsted [18, 19] fits better for this purpose.

State-of-the art studies [17, 20, 21] show that Data Vault (DV) model is designed for solving the problems of flexibility and performance, enabling maintenance of a permanent system of records ("all data, all the time"). The DV incorporates concepts from massively parallel architectures, Big Data, real-time and unstructured data, and enables agility in data model re-composition in the context of rapidly changing information requirements; these features determine our data model. As formulated in [18]: "Data Vault resolved major (flexibility and performance related) DW problems by elevating staging area into a persistent system of records". In our use-case, we do not deal with DW; nonetheless, the DV approach classified as operational is the best choice for our system. As stated in [22], it is an extension to the DV that is immediately accessed by operational systems, and is used when the real-time support as well as reading or writing directly to the data storage is required.

DV includes *hubs*, *links*, and *satellites*. We illustrate DV data structures with an example in Fig. 3a. Hubs correspond to objects of the conceptual model represented as unique lists of business keys that are used to track and identify key information (e.g. HUB_PATIENT and HUB_REHAB_PROGRAM). Links define relations between objects (e.g. LNK_PATIENT_REHAB_PROGRAM). Satellites contain descriptive attributes of the objects (e.g. SAT_PATIENT, SAT_PHYSICAL_FINDING).

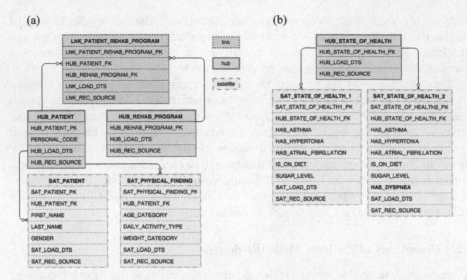

Fig. 3. (a): Representing entities *Patient* and *RehabProgram* in hub/link/satellite data structure, (b): Two versions of *StateOfHealth* satellite table (before/after adding *HasDyspnea* attribute).

Each structure element of the DV has its load date timestamp (<structure>_LOAD_DTS) and record source (<structure>_REC_SOURCE).

In our use-case, the determining factors are the flexibility of restructuring and historicity of the data. Change management should be available not only at the development stage, but also to framework users (e.g. support/non-technical staff) in the process of its operation. Hence, it leads to introducing a functional layer to describe the transformations both in data structures and mappings by means of the graphical user interface (GUI). A subject-oriented data definition language (DDL) should be provided to operate freely with conceptual objects. The concept-based statements should be translated to the 2nd layer DDL statements applied to DV objects. We are evaluating the existing solutions on the subject of their applicability to our domain of study.

Let's consider an example of representing *StateOfHealth* entity of the conceptual model (Fig. 2) as a DV object. The corresponding satellite table SAT_STATE_OF_HEALTH_1 (Fig. 3b) contains a set of attributes that got extended with an attribute HAS_DYSPHNEA represented in a new satellite table SAT_STATE_OF_HEALTH_2. Starting from the moment when the attributes are added to the existing satellite table, their values in the old records of the table will be empty. Thus, the application of a multi-instanced approach to represent satellites is more rational when creating a new satellite with a new set of attributes linked to the same hub table (HUB_STATE_OF_HEALTH). Usage of satellite tables is determined by the time of creation of a new satellite, and also by a particular treatment program. Thus, the whole history of changes in entity structure is being preserved.

The core of any rehabilitation program is a set of exercises during each stage. In the course of time the set may be altered (e.g. some exercises get eliminated/added). This way, hub tables HUB_EXERCISE and HUB_REHAB_PROGRAM should be re-linked maintaining the history of all changes. For example, if two new exercises are added to the link table, then for the rehabilitation program with the code, say, *RP1* the old exercise code *E1* remains as a historical record and two new exercise codes (exercise code is *E2* or *E3*) are linked as current ones. Data on physiotherapists that restructures a rehabilitation program is also stored in the link table.

4.3 Analytical Reports to Assess the Effectiveness of Rehabilitation Programs and Technical Implementation Features

Analytical reports available in the framework are supposed to be of various types with filtering capabilities on a set of parameters such as an arbitrary combination of attributes, time restrictions, etc., also including ad-hoc queries (e.g. rehabilitation progress of patients with a certain diagnosis in a given time period of time).

We propose to employ DV as a base data model instead of the DWH, since in our case there is no data delivered from source systems by an appropriate ETL. All stored information should be used for analytics with minimal possible structural additions. The difficulty of this approach is dictated by the specificity of the DV model, for which as far as we know, off-the-shelf applications of this kind haven't been developed yet. Thus, another task would be to create an additional – analytical – layer over DV data for reporting. Similarly to the subject-oriented DDL, it is necessary to implement an effective language for defining analytical queries, which should be translated to 2nd layer DML operating with DV objects. The susceptibility and adaptability of the reporting tool to frequent changes of both DV objects and inter-component relations must be a prerequisite. Either a DV structure or a traditional relational model to answer both operational and analytical queries would be suitable for constructing an analytical layer to store pre-aggregated data. The Starry Vault [23] approach for finding a multidimensional structure in the DV is an advanced option for this purpose.

The need to implement the subject-oriented DV DDL and analytical layer raises the issue of technical implementation of the data model. From the perspective of the language, it is hardly possible to find an alternative to SQL; nevertheless, the question is what kind of database management system (DBMS) to employ. On one hand, the developed database should function as a conventional OLTP system supplemented by subject-oriented DDL. On the other hand, high efficiency is required both in carrying out structural transformations and in performing analytical queries. For this purpose, the following new kind of DBMSs are suitable: (i) DBMS that ensure effective tools for working with columnstore indexes, or (ii) hybrid DBMSs [24] that provide capabilities of Hybrid transaction/analytical processing (HTAP), and are equally productive for OLTP and analytical tasks that include high efficiency in-memory processing of complex analytical queries. The need for turning to the DBMSs of the new breed is dictated by the fact that for both DDL operations and analytics at a low level of data access high performance of indexes is required for an arbitrary combination of DV table objects. Namely, this is a typical feature of the above-mentioned types of DBMS.

5   Conclusions and Future Work

A prototype of the currently developed system consists of wearable device and mobile software for health data acquisition. A mobile application was developed to make calculations, analyse collected data from sensor nodes, store and visualize it. The framework includes functionality for communication with a patient via notifications.

We plan to provide functionality for risk factor detection for patients undergoing the post-operative rehabilitation procedure. The system should be able to detect abnormal values in certain individuals/cohorts of patients to help physiotherapists discover the potential problems and prevent the risk of gaining a repeated trauma in the course of time. The existing prototype will be examined to study positioning of the sensor nodes for more precise harvesting of vital signs. Adding supplementary sensors such as gyroscopes and more advanced signal processing including improved data filtering and sensor fusion algorithms can also affect system performance.

In this paper, we summarize that the overall requirements for the data model that would ensure the flexibility and performance are the following – it should give maximum simplicity to the maintaining of: (i) changes in the structure of all entities, (ii) changes of inter-component relations, and (iii) history of all changes made. DV could be adapted to frequent changes in information requirements – new elements are easy to add to the model without re-design of the existing structures and data load routines.

Other subjects for future research are: (i) representation and usage of baseline patterns for each exercise in the database, (ii) remote (unattended) control over rehabilitation exercise execution, and (iii) tracing the wearable device workability in real-time. A baseline is a person-oriented pattern typically derived for a cohort of patients and is based on patients' attribute data. In a data storage, it is represented as a pattern applied to the real-time streaming data coming from embedded devices with IMU during the exercise execution. The purpose of baseline matching is to trace the correctness of patient's exercise execution in real-time and to provide notifications to help a patient perform the exercise in accordance with the rehabilitation program.

Another direction of the study is detection of bad data coming from the wearable devices. The key points of investigation are: (i) pre-configuration for tracing bad data by deriving a pattern to be applied to the data flows coming from wearable devices; (ii) bad data pattern recognition in real-time during concurrently performed exercise sessions. We envision the process of bad data detection as stream analytics with flexible reconfiguration.

References

1. American Academy of Orthopaedics: Common Knee Injuries. http://orthoinfo.aaos.org/topic.cfm?topic=a00325. Accessed 10 June 2017
2. Bergmann, J.H.M., et al.: An attachable clothing sensor system for measuring knee joint angles. Sensors 13(10), 4090–4097 (2013)
3. Pasa, L., Visna, P.: Suture of meniscus. Scr. Med. (Brno) 78(3), 135–150 (2005)

4. Kuhn, H.H., Kimbrell, W.C.J.: Electrically conductive textile materials and method for making same. Patent US4803096, USA, 7 February 1989
5. Foody, J., et al.: A prototype sourceless kinematic-feedback based video game for movement based exercise. In: Proceedings of IEEE International Conference on Engineering in Medicine and Biology Society, vol. 1, p. 5366 (2006)
6. Matsubara, M., et al.: The effectiveness of auditory biofeedback on a tracking task for ankle joint movements in rehabilitation. In: Proceedings of ISon 2013, Fraunhofer IIS, Germany, pp. 81–86 (2013)
7. Yamada, T., et al.: A stretchable carbon nanotube strain sensor for human-motion detection. Nat. Nanotechnol. 6, 296–301 (2011)
8. Guo, Y., et al.: A low-cost body inertial-sensing network for practical gait discrimination of hemiplegia patients. Telemed. E-Health 18(10), 748–754 (2012)
9. Raya, C., et al.: KneeMeasurer. A wearable interface for joint angle measurements. In: Proceedings of DRT4all, pp. 89–96 (2007)
10. Chen, K., et al.: Wearable sensor-based rehabilitation exercise assessment for knee osteoarthritis. Sensors 15(2), 4193–4211 (2015)
11. Seel, T., Raisch, J., Schauer, T.: IMU-based joint angle measurement for gait analysis. Sensors 14(4), 6891–6909 (2014)
12. Daponte, P., et al.: Design and validation of a motion-tracking system for ROM measurements in home rehabilitation. Measurement 55, 82–96 (2014)
13. Yurtman, A., Barshan, B.: Automated evaluation of physical therapy exercises using multi-template dynamic time warping on wearable sensor signals. Comput. Methods Programs Biomed. 117(2), 189–207 (2014)
14. Riaz, Q.: Human Motion Analysis Using Very Few Inertial Measurement Units. Bonn, Germany (2015). http://hss.ulb.uni-bonn.de/2016/4243/4243.pdf. Accessed 10 June 2017
15. Hermanis, A., et al.: Demo: wearable sensor system for human biomechanics monitoring. In: Proceedings of EWSN 2016, pp. 247–248. Junction Publishing, Canada (2016)
16. van der Lans, R.F.: Developing a bi-modal logical data warehouse architecture using data virtualization (2016). Denodo whitepapers http://www.denodo.com/en/document/whitepaper/developing-bimodal-logical-data-warehouse-architecture-using-data-virtualization. Accessed 10 June 2017
17. Mesaglio, M., Mingay, S.: Bimodal IT: how to be digitally agile without making a mess (2014). Gartner https://www.gartner.com/doc/2798217/bimodal-it-digitally-agile-making. Accessed 10 June 2017
18. Jovanovic, V., Bojicic, I.: Conceptual data vault model. In: Proceedings of SAIS 2012, vol. 22 (2012)
19. Linstedt, D.: Data vault basics. Dan Linstedt https://danlinstedt.com/solutions-2/data-vault-basics/. Accessed 10 June 2017
20. Inmon, W.H., Linstedt, D.: Data Architecture: A Primer for the Data Scientist - Big Data, Data Warehouse and Data Vault. Morgan Kaufmann, 378p. (2015)
21. Krneta, D., Jovanovic, V., Marjanovic, Z.: A direct approach to physical data vault design. Comput. Sci. Inf. Syst. 11(2), 569–599 (2014)
22. Linstedt, D., Olschimke, M.: Building a Scalable Data Warehouse with Data Vault 2.0. Morgan Kaufmann, 684p. (2015)
23. Golfarelli, M., Graziani, S., Rizzi, S.: Starry vault: automating multidimensional modeling from data vaults. In: Pokorný, J., Ivanović, M., Thalheim, B., Šaloun, P. (eds.) ADBIS 2016. LNCS, vol. 9809, pp. 137–151. Springer, Cham (2016). https://doi.org/10.1007/978-3-319-44039-2_10
24. Pezzini, M., et al.: Hybrid transaction/analytical processing will foster opportunities for dramatic business innovation (2014). Gartner https://www.gartner.com/doc/2657815/hybrid-transactionanalytical-processing-foster-opportunities. Accessed 10 June 2017

Towards Care Systems Using Model-Driven Adaptation and Monitoring of Autonomous Multi-clouds

Andreea Buga[✉], Sorana Tania Nemeş, and Klaus-Dieter Schewe

Johannes-Kepler-University Linz, Linz, Austria
{a.buga,t.nemes}@cdcc.faw.jku.at, kd.schewe@gmail.com

Abstract. In cloud computing, the ability to run and manage multi-cloud systems allows exploiting the peculiarities of each cloud solution and hence optimising the performance, availability, and cost of the applications. In this paper, we investigate the use case of a robotic care system as an application of autonomous multi-clouds. We present requirements and properties of an Abstract State Machines-based conceptual model that coordinates the multi-cloud interaction through the specification of a middleware exploiting adaptive interfaces to multiple clouds and supporting various service formats. While the multi-cloud system is running in normal mode, data about the execution will be gathered and evaluated by the monitoring component, and in case any critical situation is discovered the adaptation component is alerted. We show that for the care system this can be fruitfully exploited for failure alerts, failure anticipation and prevention, and safety hazards detection.

1 Introduction

The support of elderly people with special needs in their own household is an often proclaimed goal that is to offer advantages for the clients such as remaining in a familiar environment and maintaining an almost normal lifestyle as well as for caregivers who can extend services without the need for additional special buildings and specialised staff. While the privacy-critical monitoring of clients by many sensors is debatable, tailor-made care robots that accept the client as master can provide a more acceptable alternative. Such a care robot can be designed to offer a set of services to its client such as delivering small goods, reminding about necessary activities (drinking, eating, drug intake, etc.), observing safety hazards (downfall, cooking tops, household ladders, etc.) and creating corresponding alerts.

The functionality of such a care robotic system can be significantly improved, if the robot interacts with a cloud. In doing so observations including images and sound can be transmitted to cloud-based analysis services, which by means of

The research reported in this paper has been supported by the Christian-Doppler Society in the frame of the Christian-Doppler Laboratory for Client-Centric Cloud Computing.

S. de Cesare and U. Frank (Eds.): ER 2017 Workshops, LNCS 10651, pp. 26–35, 2017.
https://doi.org/10.1007/978-3-319-70625-2_3

powerful data analysis features can detect critical situations and instruct the robot about compensation actions. Furthermore, the correct functioning of the robot itself can be monitored in this way.

In this paper we present essential requirements of such a robotic care system. We then argue that the symbiosis of the care robot and the cloud-based headquarter form a typical distributed system, that can be extended to be (self-) adaptive, i.e. the robot adapts its service functionality by learning more and more about the client's real needs. Furthermore, such a system can integrate multiple robots such that also robot-to-robot interaction via the cloud can be enabled, and cloud services on multiple clouds can be exploited. For this we propose the use of the distributed, adaptive client-cloud interaction middleware (CCIM) that is presented in [6].

In general, distributed adaptive systems consist of several components that interact concurrently and are usually distributed over a network. Components are supposed to enter or leave the collection at any time; they may also be subject to change. In order to provide guarantees for the functioning of the distributed system as a whole the emphasis is therefore on adaptivity that is realized by *monitoring* components [8, 19, 20] that observe the running of the system, and *adaptation* components that change components in case of anomalies identified by the monitoring [9, 10, 13, 16].

The CCIM emphasises service-oriented distributed systems, where the software services reside on multiple clouds. Such services can be modeled as abstract state service [14] comprising a hidden internal layer and a visible and accessible view layer on top of it. The services can be integrated using a mediator model [15], which can be seen as providing general skeletons for the distributed systems that can be instantiated by concrete services that are selected according to a service ontology comprising functional, categorical and SLA-based properties. The concrete interaction of a mediator instance with the service providing clouds is subject of the middleware [2, 4], which also supports the interaction between different systems through the clouds [3].

In addition, the CCIM supports monitoring and adaptation layers [6]. That is, in normal operation mode a mediator instance will be executed, for which the middleware will be exploited to realise the interaction with the clouds and the integration of the individual services. This execution is then observed by the middleware layer, for which techniques for client-side cloud monitoring will be utilised [12]. The monitoring component further analyses the observed data in order to detect anomalies.

The remainder of this article is organised as follows. In Sect. 2 we describe the requirements of a cloud-based robotic care system emphasising structural services the robot offers to its client, operational services that concern the support of robots by the cloud, and the interaction of robots through multiple clouds. In Sect. 3 we briefly look into the architecture of the middleware with particular emphasis on monitoring and adaptation. We conclude with a brief summary and outlook in Sect. 4.

2   Requirements of a Robotic Care System

We think of a robotic care system[1] that can support elderly people in everyday life activities. While the core of the care system, the robot, acts autonomously and cooperates with its client, a key feature is its connection to a cloud to process information, increase its knowledge, and to share information with other care robots.

2.1   Structural Services

On a structural level the care robot is able to move around in the elderly person's household. It provides means for audio-visual interaction comprising a tablet screen for additional input/output of information, a camera to monitor the household and the client in order to detect critical situations (e.g. special requests of the client, reminders or alerts, etc.), a microphone for receiving requests by the client and enabling the discovery of critical situations by analysing sound signals (e.g. a cry, the sound of a falling object, etc.), and speakers to deliver sound messages. Furthermore, the robot is equipped with sensors for temperature and humidity, and with a tray attached to its body for delivering small items (e.g. drugs the client has to take).

Abstracting from the physical machine level comprising among others the electric drive, the sensors and the input/output devices we consider the core of the care system as a service-oriented system. Thus, on the structural level we propose the following list[2] of services having as a starting point previous user studies [11]:

Bring(x). When issued the robot is to bring the small item x to the client. For instance, these may be the drugs the client has to take at some time of the day. The service may be issued by the client using either the tablet or an audio command. The service may also be issued by the robot itself using the knowledge of the client's needs. The execution of the service requires to locate x, pick it up, locate the patient, move to the patient, and issue a audio and visual message using the tablet screen and the speakers that x can be taken from the tray.

Come. When issued by the client, the robot is to move to the client and to wait for further instructions. Likewise, a **Go** service requires the robot to leave to a default location.

Report($t(x)$). The service applies to a task t (with parameters x) that is executed by the robot. When such a task is started, the robot will record the kind of task, the time of issuing, the issuer, and (if applicable) any rationale for the task being issued. It will further record its activities as well as relevant activities in the environment (e.g. movement and other actions by the client)

---

[1] Our model is based on the commercially available GrowMu (see http://www.growmeup.eu/index.php/home/growmu-robot) robot, but extends it in various ways.
[2] This is merely a selection of services that can be extended.

until the task is completed and completion time and result can be recorded as well. Reporting can be either a routine service after completion of a task or in case of handling critical situations can comprise a sequence of activities involving interaction with the cloud.

Alert(x). The service applies to a critical situation (e.g. a cooking top that needs to be switched off, an object on the floor that may cause downfall of the client, a refridgerator door that needs to be closed) detected by the robot or transmitted by the cloud on grounds of information received from the robot. The service results in audio-visual information delivered to the client and involves the sending of the information to the cloud.

Observe($t(x)$). The service will record information about an activity t (with parameters x) that is performed by the client or concerns any device in the household. Each such activity has a specific start and end. The collected information will be transferred to the cloud for assessment, but can also partially be analysed by the robot itself. Depending on the nature of the task the transmission of information may be continuous or executed in specified time intervals. For instance, the activity t may be simply the movement of the client, where the objective is to guarantee safety. The activity may also be cooking or preparation of drinks or food using the refridgerator, using a household ladder(!), watching TV or listening to music. The latter two activities may give information about changed patterns concerning volume, which may indicate changes in the hearing capability of the client. The information collection may use the camera, the microphone and the other sensors. For each classified activity there is an associated list of issues that are to be discovered (e.g. the risk of downfall, if applicable, or the risk of forgetful hazardous actions).

Tacit! The service can be issued by the client to stop the transmission of observations to the cloud. This is an important privacy issue. Nonetheless, the robot may continue safety controlling tasks and terminate the non-transmission mode in case of an incident that required an alert. The mode will also be terminated after a specified time period has elapsed (to prevent that the client simply forgets about the service being requested), in which case the client will be informed. Likewise a **Mute!** service will cause the robot to be quiet, and a **Still!** service will prevent the robot from moving.

Note that all these structural services assume a proper functioning of the robot.

2.2   Operational Cloud-Supported Care Services

On an operational level the care robot is to learn the elderly person's needs and habits over time and enhance its functionality, which permits compensation for the degradation of the client's capabilities, and supports encouragement for remaining active, independent and socially involved. For this the cloud maintains an anonymised profile of the client comprising routine activities, special care needs, risks that require observation, particular interests, etc. The observations collected by the robot are subject to machine learning mechanisms that enable to learn changes to the profile. All services provided by the care robot depend on such profiles.

Profiles are to be maintained by service knowledge bases in the cloud. As will be emphasised in Sect. 3 all robot-cloud interaction will exploit a client-cloud middleware. The middleware is used to support the interaction of a robot with cloud-based services that are used in several ways:

Failure alerts. The robot will be enabled to report any detected failure (e.g. a broken wheel) via the middleware to the central service. This extends the structural services **Observe** and **Report** to the health of the robot itself. More generally, the robot may simply transmit a continuous signal captured the way it moves, so that the analysis of any failure concerning the electric drive and the wheels can be undertaken by a cloud service.

Failure compensation. In case a failure to the physical functioning of the robot has been detected an emergency alert message will be created informing a technician who can immediately organise the failure to be repaired (or the robot to be replaced). It will further trigger emergency messages to informal and formal caregivers and the client him/herself informing them that the client is no longer supported by the robot.

Anticipation of failure situations and failure prevention. The cloud service will be extended by analysing facilities on grounds of the collected monitoring data about the robot's activities. The analysis exploits common predictive maintenance techniques to anticipate potential failures, which will trigger again warning messages to a technician who can organise preventive maintenance measures. These will also apply in cases the robot does not react or anomalies in the robot's actions are detected.

Behavioural pattern detection. The monitoring of the robot's actions is extended by data analysis services as part of the cloud-based system to discover recurring behavioural patterns that permit conclusions about the elderly person's needs. These patterns will be stored centrally.

Safety hazards detection. The monitoring of the robot's actions is extended by model-driven data analysis as part of the cloud-based system to detect safety-critical situations such as objects on the floor, insufficient distance of the robot to the elderly person or insufficient knowledge about the location of the caretaker. In these cases simple adaptation means are used ranging from stopping a running operation, requesting human interaction or information from the elderly person to pre-processed alternative action plans provided by the adaptation facility provided by the cloud.

Dealing with security. The robot-middleware-cloud system will exploit identity management and access control facilities of the middleware to enable basic security mechanisms. Furthermore, all communication between robots, the middleware and the clouds is subject to anomaly detection software as part of the monitoring layer. The adaptation means in case of a security hazard (detected intrusion or anomaly) results in a switch of communication means, i.e. the robot will be served by identical services from a different cloud using a different component of the distributed middleware.

2.3   Interacting Care Robots

In order to support optimal care the cloud-based robotic care system is to support multiple interacting care robots. This addresses a learning aspect as well as a collaboration aspect. These are used in the following ways:

Dealing with uncertainty. When a robot encounters a yet unknown situation, e.g. a novel request by the caretaker, a yet unknown safety hazard or a change in the environment, the monitoring layer of the middleware will be exploited to detect such a situation. The adaptation layer of the middleware will search for a behavioural pattern that has been used by a different robot to cover such a situation and modify the robot accordingly. If no such pattern can be found, human interaction through the already implemented alert mechanisms remain the default fallback option.

Dealing with privacy. In order to protect the caretaker's privacy, behavioural patterns will be disconnected from personal information, for which the distribution of the system will be exploited.

Collaboration among robots. For the situation that multiple robots take care of a group of elderly people, e.g. in an elderly home, each robot having dedicated tasks (e.g. being company during a walk, delivering post or medicine, remind a caretaker to drink, etc.), the robots have to interact with each other and decide (by means of consensus algorithms), which robot is responsible for a particular patient. These responsibilities may change over time, for which location-based information will be required from the robot, that is analysed by the monitoring layer to anticipate the upcoming needs of the patient, and the adaptation layer will realise the change of responsibilities when needed. This will be extended further to capture failure situations, where other robots will have to step in to replace the functionality of a broken down "colleague".

3   The Multi-cloud Architecture

The extended CCIM proposed by [6] addresses the coordination of resources and the routing of client requests to individual services through a set of communicating middleware instances. The core of the model is an abstract machine defined in terms of ambient Abstract State Machines (ASM). The architecture follows a layered organization detailed in [7], in which the execution layer is supported by rollback and restart engines, communication and request handlers, service interface and a dynamic deployment. On top of the execution layer, monitoring and adaptation are defined.

The monitoring and adaptation layer of the solution run in the background of the execution processes and handle the detection and recovery from faulty situations, as for instance incorrect behavior of the robot or unresponsiveness to commands. They also focus on the reaction of the robots to safety critical situations (presence of unknown objects, hazards, etc.).

3.1 Monitoring

Monitors assess the status of a robot at any moment in time. Being part of a multi-cloud system, monitors also face possible failures. In order to avoid the problem of a single point of failure, we opted for assigning a set of monitors to each robot and to evaluate possible issues with the aid of a collaborative decision. Also, each monitor is assigned a confidence measure, which reflects its accuracy in detecting failure and safety critical situations. Monitors with low values are removed and either restarted or replaced.

We consider that a middleware component handles the assignment of monitors to each robot. The middleware is also responsible for electing a leader for each set of monitors. The leader introduces a hierarchical view to the approach and is responsible for coordinating information from different monitors. A possible formalization of the leader election algorithm is presented by [1].

The monitoring framework and the knowledge structure proposed by [5] were previously encompassed in an ASM model, which was simulated and validated. Properties of the model are now specified with the aid of Computation Tree Logic (CTL) and address aspects like safety and reachability.

First reachability property addresses the collaboration between a monitor and a leader. It implies that for all the situations in which a monitor wants to report a problem, the leader eventually moves to the *Evaluate* state. We used a static leader in this case as we could not integrate a function to bind a monitor to a leader inside the CTL formula.

CTLSPEC (forall \$m **in** Monitor **with** ag((trigger_gossip(\$m) = true)
implies ef(leader_state(leader_1) = EVALUATE)))

Listing 1.1. Reachability property of the leader agent

Monitors can identify a problem when analyzing the data they collected. For instance, the detection of an unknown object for the robot. After identifying an issue, the monitor must report it. The following property verifies this action sequence and ensures that no issue remains unreported to the leader, which triggers afterwards a collaborative diagnosis.

CTLSPEC (forall \$m **in** Monitor **with** ag((is_problem_discovered(\$m))
implies ax(monitor_state(\$m) = REPORT_PROBLEM)))

Listing 1.2. Monitor liveness property

In the case the monitor does not identify an issue, it must not reach the *REPORT_PROBLEM* state. By checking this property, the monitors are prevented to report false positives.

CTLSPEC (forall \$m **in** Monitor **with** ag(monitor_state(\$m) =
ASSIGN_DIAGNOSIS and (not(is_problem_discovered(\$m))) implies
ex(monitor_state(\$m) = LOG_DATA)))

Listing 1.3. Monitor safety property

We also verified that the leader agent starts and resets its counters for a diagnosis to zero. This property checks the fairness of a leader towards reaching a correct assessment. It is not biased towards a specific diagnosis, before inquiring the monitors, but rather it starts from the premise there is an equal chance that the observed component is in a normal, critical or failed situation.

```
CTLSPEC (forall $l in Leader with ag( (leader_state($l) = IDLE_LEADER)
    implies ax(failed_diagnoses($l) = 0 and critical_diagnoses($l) = 0 and
    normal_diagnoses($l) = 0) ))
```

Listing 1.4. Leader fairness property

The properties were tested with the AsmetaSMV Eclipse plugin[3].

3.2 Adaptation

The Adaptation Engine aims to perpetually react to the input measurements and notifications from the monitoring component and maintain its resiliency to gracefully handle and adapt to new contexts. The Adaptation Engine is comprised of two major parts, each with well delimited responsibilities and areas of inference and control.

The decision phase is defined by solution exploration, identification and maintenance [17]. Solutions are generated and adjusted to higher levels of quality compliance, taking into account the data collected and assessed by the monitoring components, the registered history of system faults and linked standardized repair actions, as well as similarity to certain key comparison factors and indicators (e.g.: availability, response time). In the solution management and enactment phase [18], any employed adaptive collaboration pattern is configured and stored in the repository as a workflow schema detailing the actions and underlying transition dependencies needed to restore the system to a normal execution mode.

Starting from the ground models we proceeded through a chain of refinements to obtain more detailed models of the adaptation components. The encoding of the ground models into ASMETA was done by using the concrete syntax AsmetaL. Next step was validating the models to check if the system behaves as expected, and if the models correctly capture the intended requirements. We were able to perform the model simulation through scenarios in both an interactive and automatic way by utilizing the simulator AsmetaS and the validator AsmetaV. A scenario is usually associated with a specific execution path and can be used for testing the state of the system after a set of transitions.

Inconsistency errors were detected at simulation time with the aid of the AsmetaS tool. For example, more than one system failure can be reported in a short time frame. And although a schema is locked while its associated solution is executed, a parallel execution of simultaneous adaptations may try to update system parts or components with different values at the same time. Triggering simultaneously multiple adaptions within the system is then supported by transaction

---

[3] A more detailed specification and results of the verification can be found at http:// cdcc.faw.jku.at/staff/abuga/esocc.zip.

specific operations where every solution is annotated with extensive knowledge on the area of inference in the system of each subsequent action, which would later on be considered in the decision phase an appropriate solution.

4   Conclusion

In this paper we described an approach to the modelling of a cloud-based robotic care system. The robot system provides services supporting an elderly client in his/her household. Each robot is supported by cloud services for the analysis of observations concerning safety of the client as well as learning of profiles. Through the cloud-based system also multiple robots are supported, an knowledge gained by one of them can be exploited for the benefit of others.

We propose a realisation on grounds of a distributed, (self-)adaptive system that is based on services supported by multiple clouds. The general model for service-oriented systems that exploit cloud-enabled services is the mediator model from [15], and the concrete interaction with the multiple clouds is realised by a middleware architecture [4]. This middleware is further extended by monitoring and adaptation layers that identify the need for a change of a mediator instantiation and provide an updated one [6].

Our research is ongoing concerning the specification of details of the monitoring and adaptation layers, refinements towards verified implementations, and in particular applying the findings to the concrete robotic care system.

We let as future work, the transformation of the model to a prototype, which can be included in the services of a multi-cloud system. The interconnection of the robots and their connection to the cloud system can be done through the CCIM [3] based on various service plots matching the elderly care requirements.

References

1. Börger, E., Stark, R.F.: Abstract State Machines: A Method for High-Level System Design and Analysis. Springer, Heidelberg (2003). doi:10.1007/978-3-642-18216-7
2. Bósa, K.: Formal modeling of mobile computing systems based on ambient Abstract State Machines. SDKB 2011. LNCS, vol. 7693, pp. 18–49. Springer, Heidelberg (2012). doi:10.1007/978-3-642-36008-4_2
3. Bósa, K.: An ambient ASM model for client-to-client interaction via cloud computing. In: Proceedings of the 8th International Conference on Software and Data Technologies (ICSOFT), pp. 459–470. SciTePress (2013)
4. Bósa, K., Holom, R.M., Vleju, M.B.: A formal model of client-cloud interaction. In: Thalheim, B., Schewe, K.D., Prinz, A., Buchberger, B. (eds.) Correct Software in Web Applications and Web Services, pp. 83–144. Springer, Cham (2014). doi:10.1007/978-3-319-17112-8_4
5. Buga, A., Nemeş, S.T.: A formal approach for failure detection in large-scale distributed systems using Abstract State Machines. In: Benslimane, D., Damiani, E., Grosky, W.I., Hameurlain, A., Sheth, A., Wagner, R.R. (eds.) DEXA 2017. LNCS, vol. 10438, pp. 505–513. Springer, Cham (2017). doi:10.1007/978-3-319-64468-4_38

6. Buga, A., Nemeş, S.T., Schewe, K.D.: Conceptual modelling of autonomous multi-cloud interaction with reflective semantics. In: Guizzardi, G., Ma, H., Mayr, H.C. (eds.) Conceptual Modeling - 36th International Conference (ER 2017). LNCS, Springer (2017, to appear)

7. Buga, A., Nemes, S.T.: Towards modeling monitoring of smart traffic services in a large-scale distributed system. In: Proceedings of the 7th International Conference on Cloud Computing and Services Science - Volume 1: CLOSER, pp. 483–490. INSTICC, ScitePress (2017)

8. Calzarossa, M., Della Vedova, M.L., Massari, L., Petcu, D., Tabash, M.I.M., Tessera, D.: Workloads in the clouds. In: Fiondella, L., Puliafito, A. (eds.) Principles of Performance and Reliability Modeling and Evaluation. Reliability Engineering. Springer, Cham (2016). doi:10.1007/978-3-319-30599-8_20

9. Calzarossa, M., Massari, L., Tessera, D.: Workload characterization: a survey revisited. ACM Comput. Surv. **48**(3), 48:1–48:43 (2016)

10. Cheng, B.H.C., de Lemor, R., Giese, H., Inverardi, P., Magee, J. (eds.): Software Engineering for Self-Adaptive Systems. Programming and Software Engineering, vol. 5525. Springer, Heidelberg (2009). doi:10.1007/978-3-642-02161-9

11. Gross, H.M., Schroeter, C., Mueller, S., Volkhardt, M., Einhorn, E., Bley, A., Martin, C., Langner, T., Merten, M.: Progress in developing a socially assistive mobile home robot companion for the elderly with mild cognitive impairment. In: 2011 IEEE/RSJ International Conference on Intelligent Robots and Systems, pp. 2430–2437, September 2011

12. Lampesberger, H., Rady, M.: Monitoring of client-cloud interaction. In: Thalheim, B., Schewe, K.D., Prinz, A., Buchberger, B. (eds.) Correct Software in Web Applications and Web Services, pp. 177–228. Springer, Cham (2015). doi:10. 1007/978-3-319-17112-8_6

13. Leucker, M., Schallhart, C.: A brief account of runtime verification. J. Logic Algebr. Program. **78**(5), 293–303 (2009)

14. Ma, H., Schewe, K.D., Thalheim, B., Wang, Q.: A theory of data-intensive software services. SOCA **3**(4), 263–283 (2009)

15. Ma, H., Schewe, K.D., Thalheim, B., Wang, Q.: A formal model for the interoperability of service clouds. SOCA **6**(3), 189–205 (2012)

16. Mirandola, R., Potena, P., Scandurra, P.: An optimization process for adaptation space exploration of service-oriented applications. In: Proceedings of the 6th IEEE International Symposium on Service-Oriented System Engineering (SOSE 2011), pp. 146–151. IEEE (2011)

17. Nemeş, S.T., Buga, A.: Towards a case-based reasoning approach to dynamic adaptation for large-scale distributed systems. In: Aha, D.W., Lieber, J. (eds.) ICCBR 2017. LNCS (LNAI), vol. 10339, pp. 257–271. Springer, Cham (2017). doi:10.1007/978-3-319-61030-6_18

18. Nemes, S.T., Buga, A.: Towards modeling adaptation services for large-scale distributed systems with abstract state machines. In: Shishkov, B. (ed.) Business Modeling and Software Design - 7th International Symposium, Proceedings, BMSD 2017, Barcelona, Spain, 3–5 July 2017, pp. 193–198. Springer (2017)

19. Shin, K.S., Jung, J.H., Cheon, J.Y., Choi, S.B.: Real-time network monitoring scheme based on SNMP for dynamic information. J. Netw. Comput. Appl. **30**(1), 331–353 (2007)

20. Zeng, W., Wang, Y.: Design and implementation of server monitoring system based on SNMP. In: JCAI, pp. 680–682 (2009)

Clustering Event Traces by Behavioral Similarity

Agnes Koschmider[(✉)]

Institute AIFB, Karlsruhe Institute of Technology, Karlsruhe, Germany
agnes.koschmider@kit.edu

Abstract. The automated analysis of event logs in smart homes could provide an IT-aided support for a highly autonomous and age-appropriate life standard. The analysis of human behavior in the context of smart homes is, however, a challenging task. Humans behave according to the best practices and a single behavioral model is typically not sufficient to represent them all. In fact, existing process mining algorithms reportedly generate spaghetti models from event logs of flexible processes, which are largely incomprehensible. Therefore, this paper presents a novel approach for clustering event traces by their behavioral similarity, rather than deriving a unique process model encompassing all traces. In order to do this, two algorithms are introduced and we report the results of a preliminary evaluation demonstrating the efficacy of the approach.

1 Introduction

One of the main objectives of Ambient Assisted Living is to provide IT-aided support for a highly customized, autonomous, and age-appropriate life standard for the users of smart homes. The automated analysis of event logs reporting the inhabitants' behavior are of great help in this regard. The analysis of human behavior in the context of smart homes is, however, a challenging task. Everyday sequences of tasks are mostly based upon ad hoc, highly variable goal-oriented processes, which means that a single behavioral model is typically not sufficient to represent them all. Furthermore, a classification into good and bad behavior is largely impossible. For instance, a person who leaves home directly after breakfast instead of first watching tv, does not directly show abnormal behavior. In fact, existing process mining algorithms reportedly generate spaghetti models from event logs, which are largely incomprehensible. This calls for techniques which *identify* and *explain* behavioral deviations (i.e., explain in detail why something is going to happen).

This paper provides a novel approach for clustering event traces by their behavioral similarity. The technique can compare the control flow of event traces and relate this comparison to time, duration, and further exogenous factors such as temperature. The improved classification of event traces makes our technique superior to existing approaches, allowing more efficient identifying and analyzing behavioral deviations. Our approach is based on smart home sensors which transmit event streams (e.g., sensor data), which must first be splitted up

© Springer International Publishing AG 2017
S. de Cesare and U. Frank (Eds.): ER 2017 Workshops, LNCS 10651, pp. 36–42, 2017.
https://doi.org/10.1007/978-3-319-70625-2_4

according to entities (e.g., patient, nurse, relative) resulting in an entity-centric trace. This step is not addressed in this paper but we assume that the data cleansing has been performed. Based upon an entity-centric trace, the clustering technique builds on two algorithms as explained in Sect. 3. A behavioral morphing algorithm determines all valid behavior states and represents each state as a so-called behavior-oriented trace. The assignment algorithm determines the similarity between behaviors. Both algorithms lay the foundation for novel clustering of event traces as described in Sect. 4. The comparison of our clustering technique to existing clustering approaches is presented in Sect. 5. Section 6 concludes the paper.

2 From Event Logs to Behavior-Oriented Traces

The input of the behavioral morphing and assignment algorithms is a so-called behavior-oriented trace that is a textual representation of a process tree of the form $<\rightarrow(a,\wedge(b,c),\wedge(d,e))>$. A process tree [1] can be seen as an extension of regular expressions and it allows for defining block-structured process models in a recursive way. Process trees rely on traces, which is a sequence of actions, and event logs, which is a multiset of traces [2]. Event logs are the fundamental input of automated process discovery algorithms that intend to generate a process model out of it. Actions in the context of Ambient Assisted Living are enter room (a), watch TV (b), prepare dinner (c), eat (d), or make call (e). A behavior-oriented trace $<\rightarrow(a,\wedge(b,c),\wedge(d,e))>$ might represent the behavior of these actions where $\rightarrow$ indicates a sequence and $\wedge$ a parallel activity execution. In the following, we represent actions as letters and we assume that cleansing techniques for action labels in terms of linguistics have been performed [3,4]. The PTandLogGenerator [5] tool can be used to derive a process tree from a graphical process model and then a transformation to behavior-oriented traces can be applied.

3 Behavioral Morphing and Assignment as Preliminaries to Clustering

This section describes the behavioral morphing and the assignment algorithms as preliminary steps for clustering of event traces.

3.1 Behavioral Morphing Algorithm

The morphing algorithm takes as input two behavior-oriented traces and represents each difference between them as an own behavior-oriented trace. The idea of the algorithm is based on state morphing as it is known from face morphing. The representation of all behavior states allows valid and possible behaviors to be understood. Basically, the first input for the morphing algorithm is the as-is behavior BT_{as} and the to-be behavior BT_{tb}. The as-is behavior can either be given as a graphical business process model (e.g., BPMN, Petri net) or as an empty trace when the as-is behavior is not known. An empty trace for the to-be behavior is not allowed and can be derived from specifications. The morphing from as-is to to-be behaviors is called a root axis. This axis can be considered as a classification schema for behaviors, which also groups event traces. The algorithm then processes any input behavior and relates it to the root axis based on the assignment algorithm explained in the next section. Generally, we consider a behavior as a 24 h/day, which we recommend to be divided into times of day (e.g., noon, afternoon, night). The morphing algorithm uses the operations *insertion*, *deletion*, and *replacement*. For this, two behavior-oriented traces, BT_1 and BT_2 are given and an alphabet $\mathcal{A} \cup \{-\}$ where $\{-\}$ denotes a gap. Additionally, let the behavioral operators $O = <\rightarrow, x, \wedge, loop>$ be given with the semantics of process tree operators. For a,b $\in \mathcal{A}$ the following edit operations are possible:

1. (a,a) means a match of symbols between BT_1 and BT_2 at some position of BT_{1_i} and BT_{1_j}.
2. (-,(a|O)) denotes an insertion of a or any behavioral operator in BT_2. Insertion is used if additional activities or operators are given in BT_2 and miss in BT_1.
3. ((a|O),-) denotes a deletion of a or any behavioral operator from BT_1.
4. ((a| O), (b|O)) means a replacement of a from BT_{1_i} by b from BT_{2_j} where a $\neq$ b or replacement of behavioral operators (e.g., see Fig. 1 where the alternative operator is replaced by the sequence operator).

In Fig. 1, the morphing algorithm is applied for the as-is behavior $<\rightarrow(\wedge(a,b),$ $x(c,d))>$ and an example behavior $<\wedge(b,x(c,d))>$. First, the algorithm is applied for the as-is to the to-be behavior (representing the root axis). In this figure, the graphical representation of the process tree is used to better visualize the

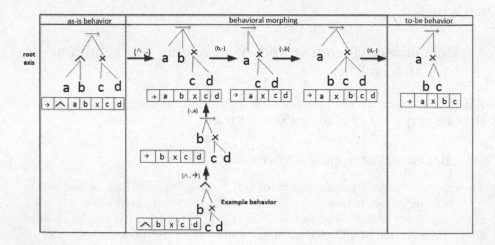

Fig. 1. Morphing of three behavior states.

novelty of our approach, namely the grouping of behaviors. To know where to morph the behavior $<\wedge(b,x(c,d))>$ first requires the identification of its closest behavior based on the assignment algorithm that follows.

3.2   Assignment Algorithm Based on Edit Distance

The assignment algorithm computes the edit distance between two behavior-oriented traces and intends to find the shortest sequence of edits between them. Generally, the assignment algorithm works similarly to the Levenshtein distance that intends to find the shortest sequence of edits required to transform one string into another. Let S be a sequence of edit operations $s_1, ..., s_k$ given by $\gamma(S) = \sum_{i=1}^{k}(s_i)$. Then we define the edit distance between two behavior-oriented traces BT_1 and BT_2 by: $\delta(BT_1, BT_2) = min\{(S)$—$S$ is a sequence of edit operations transforming BT_1 to $BT_2\}$. For illustration let the behavior-oriented trace $T_b=<\wedge,b,x(c,d)>$ be given. Then, the assignment algorithm searches for the behavior with the shortest sequence of edits to T_b. It compares each behavior of the root axis (see Fig. 2), which matches the second behavior $<\rightarrow(b,x(c,d))>$. In case of identical sequence of edits to an input behavior, the behavior-oriented trace with the lowest number of replacement operators is selected. This ensures that the behaviors are relatively close to each other. If the number of replacement operators is identical, then the intersection of activities is considered.

Fig. 2. Application of the alignment algorithms for behavior $<\wedge(b,x(c,d))>$.

4   Clustering Technique and Analysis

The behavioral morphing and assignment algorithms lay the foundation for clustering aimed at an improved analysis of behavioral deviations. In our approach, behavioral deviations are detected by a simulation of clusters by means of time. We define a cluster C by (R, P) where R is the root and P is a set of vertical or/and horizontal paths of behavior-oriented traces to the root. The size of the cluster (vertical path vs. vertical and horizontal paths) depends on the focus of analysis. For instance, the behavior-oriented trace $<\wedge,b,x(c,d)>$ (see Fig. 1) belongs to the (vertical) cluster with the behaviors $<\rightarrow(b,\dot{x}(c,d))>$ and $<\rightarrow(a,b,x(c,d))>$ where the latter is the root of this cluster. We are interested in any deviations in this cluster. For cluster analysis, the frequency for each behavior-oriented trace (i.e., occurrence of a behavior) is counted and denoted by ($\#$). Each occurrence of a behavior is plotted to a group and compared with respect to exogenous factors (e.g., duration, temperature) and by means of behavioral operators (alternative, parallel) giving hints to an increased duration of activities or correlation to exogenous factors. The left column of Fig. 3 shows the cluster analysis according to occurrence (#) and exogenous factors. For instance, the behaviors P_{G2} and P_{G21} (see Fig. 3a) occurred almost equally. When relating these two behaviors to time, differences with respect to behavioral operators and activities can be identified (see Fig. 3b). For instance, activities C and D vary according to time. Also, activity A only occurred in the behavior P_G. To rate the importance of activity A, the clusters are plotted by the number of occurrence and represented graphically (see Fig. 3c). The size of the graphical clusters correlates with the number of occurrences. When simulating the clusters over time, an increasing size means a steady occurrence of a behavior and indicates shifts between current versus a past behavior. Activity A occurs in P_G and P_{G3}, but both behaviors occurred recently rarely.

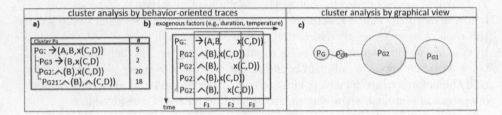

Fig. 3. Two representations for cluster analysis

5   Evaluation

We conducted a preliminary evaluation of our approach and compared related clustering approaches. Generally, the approaches rely on bag-of-words [6] or edit distance [7,8]. We compared the approaches with respect to the identification of behavioral deviations in the context of smart homes[1] that are (1) changes of day and night activities (e.g., increased number of activities at night, leaving home at unusual times), (2) changes of activities during the day (e.g., increased duration for resting or sleeping or changes in the number activities), and (3) change of common (process) activities (e.g., habits changed by means of time outside the home). All these changes need to be related to exogenous factors. The comparison shows that our approach is superior with respect to behavioral deviations (see Table 1) and can be explained by the precise clustering of behaviors based on their behavioral similarity. Related approaches can only identify missing activities or deviations of common activities. Our cluster analysis allows to identify (1): changes of day and night activities through the graphical cluster analysis, (2) changes of activities during the day through the comparison of

---

[1] These requirements generally define influences on cognitive impairments.

behavior-oriented traces against durations. (3) changes of common activities by a time-oriented comparison of behavior-oriented traces. (4) Behavior-oriented traces can be related to duration and any other exogenous factor.

Table 1. Comparison of related clustering approaches

Related approach	Change day, night	Change activities	Change common activities	Exogenous factors
Song [6]	x	-	-	-
Bose [7]	x	-	-	-
Hompes [8]	-	x	x	-
Koschmider	x	x	x	x

6 Conclusion and Outlook

This paper presented an approach for clustering event traces based on their behavioral similarity, rather than deriving a unique process model that encompasses all traces. The clustering technique is based on two algorithms which classify behaviors and allow us to identify behavioral shifts. The only input required for our approach is the to-be behavior, which might be derived from specifications. In contrast to machine-learning techniques, no large input sample is required. Our approach brings transparency for machine-learning techniques allowing us to explain *why* behavior will deviate.

References

1. Buijs, J.C.A.M., van Dongen, B.F., van der Aalst, W.M.P.: A genetic algorithm for discovering process trees. In: IEEE Congress on Evolutionary Computation, pp. 1–8. IEEE (2012)
2. Fahland, D., van der Aalst, W.M.: Model repair - aligning process models to reality. Inf. Syst. **47**, 220–243 (2015)
3. Ehrig, M., Koschmider, A., Oberweis, A.: Measuring similarity between semantic business process models. In: APCCM. CRPIT, Vol. 67, pp. 71–80. Australian Computer Society (2007)
4. Koschmider, A., Ullrich, M., Heine, A., Oberweis, A.: Revising the vocabulary of business process element labels. In: Zdravkovic, J., Kirikova, M., Johannesson, P. (eds.) CAiSE 2015. LNCS, vol. 9097, pp. 69–83. Springer, Cham (2015). https://doi.org/10.1007/978-3-319-19069-3_5
5. Jouck, T., Depaire, B.: PTandLoggenerator: A generator for artificial event data. In: Proceedings of the BPM Demo Track 2016, CEUR-WS.org, pp. 23–27 (2016)
6. Song, M., Günther, C.W., van der Aalst, W.M.P.: Trace clustering in process mining. In: Ardagna, D., Mecella, M., Yang, J. (eds.) BPM 2008. LNBIP, vol. 17, pp. 109–120. Springer, Heidelberg (2009). https://doi.org/10.1007/978-3-642-00328-8_11

7. Bose, R.P.J.C., van der Aalst, W.M.P.: Context aware trace clustering: towards improving process mining results. In: Proceedings of the SIAM International Conference on Data Mining (SDM 2009), pp. 401–412 (2009)
8. Hompes, B., Buijs, J.C.A.M., van der Aalst, W.M.P., Dixit, P., Buurman, H.: Detecting change in processes using comparative trace clustering. In: Proceedings of the 5th International Symposium on Data-driven Process Discovery and Analysis (SIMPDA 2015), Vienna, Austria, 9–11 December 2015, pp. 95–108 (2015)

MoBiD 2017 - 6th International Workshop on Modeling and Management of Big Data

Preface

David Gil[1], Thomas Bäck[2,3], Nicholas Multari[4], Juan Trujillo[5],
Heike Trautmann[6], Jesús Peral[5], Gottfried Vossen[6], and Il-Yeol Song[7]

[1] Lucentia Research Group, Department of Computing Technology and Data
Processing, University of Alicante, Alicante, Spain
dgil@dtic.ua.es
[2] Leiden University, Leiden, The Netherlands
T.H.W.Baeck@liacs.leidenuniv.nl
[3] Divis Intelligent solutions GmbH, Dortmund, Germany
[4] Pacific Northwest National Laboratory, Richland, WA 99352, USA
nick.multari@pnnl.gov
[5] Lucentia Research Group, Department of Software and Computing Systems,
University of Alicante, Alicante, Spain
{jtrujillo, jperal}@dlsi.ua.es
[6] ERCIS, University of Münster, Münster, Germany
trautmann@wi.uni-muenster.de, g.v@wwu.de
[7] College of Computing and Informatics, Drexel University Philadelphia,
Philadelphia, PA 19104, USA
song@drexel.edu

In the last decade, technology has advanced tremendously. Currently, a wide variety of devices, including sensor-enabled smart devices, and all types of wearables, connect to the Internet and power newly connected applications and solutions. On the one hand, the cost of technology has sharply decreased, making it possible for everybody to engage in sensing data. The vast amount of real time information can be accessed across the Internet. Furthermore, some of the environments are just online, like social media, where all the information is in the Cloud. As a result, new words as well as new expressions have appeared such as Big Data, Cloud Computing or Internet of Things, among others.

Due to all of these enormous amounts of data generated, there is an increasing interest in incorporating them, usually referred to as Big Data, into traditional applications. This new era of Big Data requires conceptualization and methods to effectively manage big data and accomplish intended business goals. Thus, the objective of MoBiD'17 is to be an international forum for exchanging ideas on the latest and best proposals for modeling and managing big data in this new data-drive paradigm. The workshop is a forum for researchers and practitioners who are interested in the different facets related to the use of the conceptual modeling approaches for the development of next generation applications based on Big Data.

The workshop has been announced in the main announcement venues and attracted papers from seven different countries distributed all over the world: Italy, Sweden, France, Poland, Austria, Germany, and Spain. We have finally received 8 papers and the Program Committee has selected 4 papers, making an acceptance rate of 50%.

The first paper by Akoka et al. presents an MDA approach dedicated to modeling issues of NoSQL property graph databases, encompassing the four V's. The proposal consists mainly of two steps: (1) an MDA-based forward engineering approach enabling the development of conceptual, logical, and physical models; (2) the generation of the physical database automatically generates a test database whose size is tuned using volume information. The second paper by Carnein et al. shows the omni-channel Customer Relationship Management (CRM) framework which includes data integration, data presentation and data analytics. The paper introduces three case studies which demonstrate the value that omni-channel CRM can provide to customers and companies. In the third paper, Carnein et al. propose a new stream clustering algorithm for text streams. The algorithm combines concepts from stream clustering algorithms (where new observations are either added to their closest cluster or used to initialize new clusters) and text analysis in order to incrementally maintain a number of text droplets that represent topics within the stream. The last paper by Golfarelli et al. presents an approach to automate the translation of the objectives declared by the user for visualizing the results of big data analytics into a set of most suitable visualization types. The approach enables users to specify a value for different visualization coordinates, assigns a qualitative suitability score to each visualization type, then computes a skyline-based technique for automatically translating a visualization context into a set of suitable visualization types. Finally, the proposal was evaluated on a real use case.

Goal-Based Selection of Visual Representations for Big Data Analytics

Matteo Golfarelli[1,2], Tommaso Pirini[1], and Stefano Rizzi[1,2(✉)]

[1] DISI, University of Bologna, V.le Risorgimento 2, 40136 Bologna, Italy
{matteo.golfarelli,tommaso.pirini,stefano.rizzi}@unibo.it
[2] CINI, Via Salaria 113, 00198 Roma, Italy

Abstract. The H2020 TOREADOR Project adopts a model-driven architecture to streamline big data analytics and make it widely available to companies as a service. Our work in this context focuses on visualization, in particular on how to automate the translation of the visualization objectives declared by the user into a suitable visualization type. To this end we first define a visualization context based on seven prioritizable coordinates for assessing the user's objectives and describing the data to be visualized; then we propose a skyline-based technique for automatically translating a visualization context into a set of suitable visualization types. Finally, we evaluate our approach on a real use case excerpted from the pilot applications of TOREADOR.

Keywords: Big data · Visual analytics · Skyline queries

1 Introduction

As a consequence of the wide diffusion of big data technologies and of the increasing amounts of valuable data generated by sensors, devices, social media, etc., companies of all sizes have become aware of the opportunities lying with *big data analytics* (BDA), where advanced analytic techniques operate on big data sets aimed at complementing the role of traditional OLAP and data warehouses [15]. However, the lack of in-house technical skills often prevents companies from really benefiting of BDA, or even discourages them from taking this direction because of the outsourcing costs. In this context, the H2020 TOREADOR (TrustwOrthy model-awaRE Analytics Data platfORm) Project adopts a *model-driven architecture* (MDA [11]) to streamline BDA processes and make them widely and easily available to companies following a BDA-as-a-service approach. Following the basic principles of MDAs, TOREADOR builds on three models to support BDA [2]:

1. **Declarative Model**: an abstract and platform-independent model that specifies the user goals (*what* BDA should achieve) in terms of data collection,

---

This work was partly supported by the EU-funded project TOREADOR (contract n. H2020-688797).

S. de Cesare and U. Frank (Eds.): ER 2017 Workshops, LNCS 10651, pp. 47–57, 2017.
https://doi.org/10.1007/978-3-319-70625-2_5

preparation, analysis, and visualization. It corresponds to the *computation-independent model* in MDA terminology.

2. **Procedural Model**: a platform-neutral, vendor-independent model that specifies the algorithms for data preparation and for parallelizing and executing the analytics, as well as the way to present the results to users (*how* BDA should work). It corresponds to the *platform-independent model* in MDA terminology.

3. **Deployment Model**: the computational components and other resources for the process on a specific target execution platform (e.g., Hadoop-as-a-service). It corresponds to the *platform-specific model* in MDA terminology.

Remarkably, as required by the MDA paradigm, each model is (semi-) automatically derived from the previous one.

As sketched in Fig. 1, within the TOREADOR framework the three models are grouped into five conceptual areas: *preparation, representation, analytics, processing,* and *visualization.* The focus of this paper is on the visualization area, in particular on (i) how to specify the users objectives and describe the dataset to be visualized within the declarative model (e.g., comparison-oriented visualization of 4-dimensional numerical data with low-cardinality domains), and (ii) how to translate this specification into a concrete platform-independent solution (e.g., bar chart) within the procedural model, which will be eventually translated into a deployment model on the target execution platform (e.g., stacked-to-group bar chart in the D3 Java library, `d3js.org`). Specifically, the main contributions of this paper are:

- As part of the declarative model we define a *visualization context* based on seven prioritizable coordinates for assessing the user's objectives and conceptually describing the data to be visualized (Sect. 3).
- We describe a skyline-based technique for automatically translating a visualization context from the declarative model onto the procedural model in the form of a set of suitable visualization types (Sect. 4).

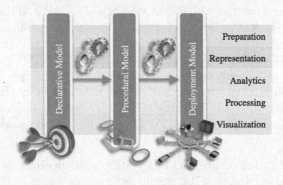

Fig. 1. The TOREADOR framework

The paper outline is completed by Sect. 2, which discusses the basic related literature, and by Sect. 5, which evaluates our approach through a real use case excerpted from the pilot applications of TOREADOR and draws the conclusions.

2 Related Work

Visualization has a key role in BDA to enable users to understand the problem, generate hypotheses and define the solution, as well as to steer the analysis process in dealing with massive, incomplete, and incorrect data [9].

Several papers propose principles and taxonomies to classify the different approaches to visualizing data and interacting with them. A seminal paper in this field is the one by Shneiderman, who proposes a classification taxonomy for data visualization based on two coordinates: *task* (e.g., overview, zoom, and details-on-demand) and *data type* (e.g., multi-dimensional, tree, and temporal) [16]. Another influential paper is [10], where Keim proposes a different classification of data visualization and visual data mining techniques by considering, besides the data type, the *visualization technique* and the *interaction and distortion technique*. A few years later, Abela listed four possible *goals* for visualization, namely relationship, comparison, distribution, and composition [1].

More recently, Börner surveyed the main classifications proposed in the literature and made a significant effort to integrate them into a single, consistent framework [4]. Her visualization framework is based on six coordinates, namely *insight need type* (which integrates [7,18]), *data scale type* (based on [17]), *visualization type* (based on [3,16]), *graphical symbol type*, *graphical variable type*, and *interaction type* (which integrates [10,16]). A more detailed classification of data types, including for instance datetime components and IRIs, is introduced in [14] with reference to the visualization of linked open data; the paper also relates each common type of chart to the user goals it is most compliant with. Finally, in [6] a new coordinate is introduced to visualize linked open data: the *user type*. Users are distinguished into *lay-users* and *techies*.

Despite the richness and detail of the classifications available, to the best of our knowledge only few papers focus on the criteria for deciding which type of chart is best suited for a given combination of data type, dimensionality, user goal, etc. In [1], a simple decision tree is proposed to select the best visualization according to the user's goal and to the main features of data (namely, the number of variables, the cyclicality, and the size). A description of the pros and cons of different charts to be used in the security domain is provided in [12]; the specific aspects of data considered include their dimensionality, cardinality, and type. A flow-chart is also provided to help users in choosing the right visualization for different goals and data dimensionality, but not all combinations are taken into account. In the context of big data, a framework for choosing the best visualization is outlined in [5]; specifically, the main types of charts are related to the user goals they fulfill and to the data dimensionality, cardinality, and type they support.

3   A Declarative Model for Visualization

In this section we describe the coordinates we use to enable users to declare their objectives and describe the dataset to be visualized. The method we followed to select these coordinates can be summarized as follows:

1. We analyzed the literature on the taxonomies of data visualization and inter-action paradigms to derive a set of candidate *coordinates* (e.g., data type) and, for each coordinate, a set of candidate *values* (e.g., ordinal).
2. From these candidate coordinates/values we derived a set of questions to be submitted to users for requirement elicitation.
3. Based on the elicitation, we selected a final set of coordinates and values.

For requirement elicitation we adopted the Kano model [8], a useful tool for understanding needs and expectations of a stakeholder based on how they affect his/her satisfaction with a given product. The Kano model classifies requirements into *must-be, one-dimensional, attractive, indifferent*, and *reverse* based on their location along two dimensions, namely, the degree of satisfaction and the level of functionality. To position each requirement a questionnaire is submitted to each user (in our context, the key users of the pilot applications of TOREADOR), then the results are aggregated and evaluated. In the following we list the coordinates we selected, see Table 1 for the values each coordinate can take:

(1) *Goal*, which enables users to declare their analysis goal. This classification follows the one into *basic task types* proposed in [4].
(2) *Interaction*, which enables users to declare the type of interactions to be sup-ported by the visualization. This classification derives from the one proposed in [4]; specifically, based on requirement elicitation, we selected a subset of most common and intuitive interaction types out of those proposed in [10].
(3) *User*, which enables users to declare their skill as in [6].
(4) *Dimensionality*, which enables users to declare the number of variables they wish to visualize. Here, as done in [1], we count all variables without distin-guishing between independent and dependent variables.
(5) *Cardinality*, which enables users to qualitatively declare the cardinality of the data to be visualized like in [1].
(6) *Type*, which enables users to declare the type of each variable to be ana-lyzed. The classification we adopt here is the one in [17], but as in [12] we distinguish independent from dependent variables.

Definition 1 (Visualization Context). *Let* $O_1, \ldots, O_7$ *be the sets of goals, interactions, users, dimensionalities, cardinalities, independent types, and depen-dent types, respectively, as listed in Table 1* ($O_6 \equiv O_7$)*; let* $C = \{1, \ldots, 7\}$ *and* $O = \bigcup_{i \in C} O_i$*. A visualization context is defined by a function* $c : C \rightarrow O \cup \{NULL\}$ *(where* $c(i) \in O_i \cup \{NULL\}$*) and by a weak order* $\overset{c}{\succ}$ *on the set* C *that expresses the priorities between the seven coordinates.*

Table 1. Visualization coordinates

Value	Objective	Example
Goal		
Composition	Highlighting the way in which distinct parts of data are composed to form a total	Stacked column chart
Order	Analyzing objects by emphasizing their ordering	Alphabetical list of names
Relationship	Analyzing the correlation between two or more objects or attribute values	Scatter plot
Comparison	Examining two or more objects or values to establish their similarities and dissimilarities	Column chart
Cluster	Analyzing data in such a way as to emphasize their grouping into categories	Dendrogram
Distribution	Analyzing how objects are dispersed in space	Histogram
Trend	Examining a general tendency of data variables	Line chart
Geospatial	Analyzing data values using a geographical map as a graphical context	Choropleth map
Interaction		
Overview	Gain an overview of the entire data collection	Dendrogram
Zoom	Focus on items of interest	Network map
Filter	Quickly focus on interesting items by eliminating unwanted items	Area chart
Details-on-demand	Select an item and get its details	Choropleth map
User		
Lay	Computer-literates who may have troubles in understanding complex visualizations	Line chart
Tech	Skilled users with a deeper understanding of BDA	Tree map
Dimensionality		
1-dimensional	A single numerical value or a string	Gauge
2-dimensional	One dependent variable as a function of one independent variable	Single-line chart
n-dimensional	Each data object is a point in an n-dimensional space	Bubble chart
Tree	A collection of items, each having a link to one parent item	Dendrogram
Graph	A collection of items, each linked to an arbitrary number of other items	Network map
Cardinality		
Low	From a few items to a few dozens items	Pie chart
High	Some dozens items or more	Heat map
Type		
Nominal	Qualitative, each data variable is assigned to one category	Pie chart
Ordinal	Qualitative, categories can be sorted	Histogram
Interval	Quantitative, it supports the determination of equality of intervals or differences	Line chart
Ratio	Quantitative, with a unique and non-arbitrary zero point	Scatter plot

Example 1. An example of visualization context is

$$c(1) = \text{Comparison}, \quad c(2) = \text{NULL}, \quad c(3) = \text{Tech},$$
$$c(4) = \text{n-dimensional}, \quad c(5) = \text{High}, \quad c(6) = \text{Interval}, \quad c(7) = \text{Ratio}$$
$$(3 \overset{c}{\sim} 6) \overset{c}{\succ} 1 \overset{c}{\succ} (2 \overset{c}{\sim} 4 \overset{c}{\sim} 5 \overset{c}{\sim} 7)$$

where the user expresses three levels of priority: high (for the user and independent type coordinates), medium (for the goal coordinate), and low (for the remaining coordinates). □

4   Going Procedural

To translate the visualization context stated by the user in the declarative model into a set of suitable visualization types in the procedural model, we first need to assess to which extent each visualization type is suitable for each value of each coordinate introduced in Sect. 3.

Definition 2 (Suitability Function). *A* suitability function *is a function* $\sigma : O \times V \to s$ *where* O *is the set of all coordinate values,* V *is the set of all visualization types, and* $s \in \{\text{unfit}, \text{discouraged}, \text{neutral}, \text{acceptable}, \text{fit}\}$ *is a suitability score.*

Our approach is general enough to be applicable to each possible visualization type v as long as a suitability evaluation is done for v based on our seven coordinates. Currently we consider a set of 25 widely used visualization types classified as shown below [4]:

- **Tables** are ordered arrangements of rows and columns in a grid, with data values stored in cells (e.g., pivot table —Fig. 2a).
- **Charts** visually depict quantitative and qualitative data without using a well defined reference system (e.g., tag cloud —Fig. 2b).
- **Graphs** plot quantitative and qualitative data using a well-defined reference system, such as Cartesian coordinates (e.g., bubble chart —Fig. 2c).
- **Maps** display data according to their spatial relationships and show how data are distributed geographically (e.g., heat map —Fig. 2d).
- **Network layouts** use nodes to represent sets of data records, and inter-node connections to represent relationships (e.g., dendrogram —Fig. 2e).

Then we defined a suitability function by assigning a score to each visualization type/coordinate value pair; the scores were derived from the literature (mostly from [1,4,12]). For instance, in Table 2 we show the suitability scores for three popular visualization types.

The next problem is that of using the suitability function to find, given a visualization context c, one or more "most suitable" visualization types. To this end we start by observing that, with reference to c, visualization type v

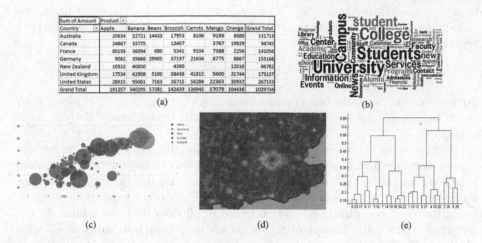

Fig. 2. A pivot table (a), a tag cloud (b), a bubble chart (c), a heat map (d), and a dendrogram (e)

Table 2. Suitability scores for three visualization types

		pie chart	bubble chart	heat map
Goal:	Composition	fit	unfit	unfit
	Order	neutral	unfit	unfit
	Relationship	unfit	fit	unfit
	Comparison	neutral	fit	acceptable
	Cluster	unfit	acceptable	acceptable
	Distribution	discouraged	fit	fit
	Trend	unfit	fit	unfit
	Geospatial	unfit	unfit	fit
Interaction:	Overview	fit	acceptable	fit
	Zoom	unfit	acceptable	fit
	Filter	neutral	neutral	neutral
	Details-on-dem	acceptable	neutral	neutral
User:	Lay	fit	acceptable	acceptable
	Tech	acceptable	fit	fit
Dimensionality:	1-dimensional	unfit	unfit	unfit
	2-dimensional	fit	unfit	unfit
	n-dimensional	unfit	fit	fit
	Tree	unfit	unfit	unfit
	Graph	unfit	unfit	unfit
Cardinality:	Low	fit	acceptable	acceptable
	High	discouraged	discouraged	fit
Independent Type:	Nominal	fit	unfit	neutral
	Ordinal	acceptable	neutral	acceptable
	Interval	discouraged	fit	fit
	Ratio	discouraged	fit	fit
Dependent Type:	Nominal	unfit	fit	unfit
	Ordinal	unfit	fit	discouraged
	Interval	unfit	fit	fit
	Ratio	fit	fit	fit

is evaluated through a 7-tuple $\langle \sigma(c(1),v),\dots,\sigma(c(7),v)\rangle$ where each element expresses the suitability of v for c along one coordinate. On the other hand, the suitability scores introduced in Definition 2 are obviously related by a (strict) total order expressing a preference:

$$\text{fit} > \text{acceptable} > \text{neutral} > \text{discouraged} > \text{unfit}$$

So we can compare any two possible visualization types $v, v' \in V$ along each single coordinate: for the i-th coordinate, v is strictly better than v' if $\sigma(c(i),v) > \sigma(c(i),v')$.

Now, we have to combine the seven resulting one-dimensional preferences into a composite one for the whole 7-tuple. A popular way to cope with this problem is to look for tuples (corresponding in our case to visualization types) that are Pareto-optimal. A tuple is *Pareto-optimal* when no other tuple dominates it, being better in one dimension and no worse in all the other dimensions. In the database community, the set of tuples satisfying Pareto-optimality is called a *skyline* [13]. The definition of dominance is given below in flat (non-prioritized) form; it is given with reference to a subset of coordinate C' to be more easily generalized to the prioritized case in Definition 4.

Definition 3 (Flat Dominance). *Given visualization context c and two visualization types v and v', and given the set of coordinates $C' \subseteq C$, we say that v is* flat-substitutable *to v' on C', denoted $v \sim_{C'} v'$, iff $\sigma(c(j),v) = \sigma(c(j),v')$ for all $j \in C'$ such that $c(j) \neq$ NULL. We say that v flat-dominates v' on C', denoted $v \succ_{C'} v'$, iff (a) $\exists i \in C' : \sigma(c(i),v) > \sigma(c(i),v')$ and (b) for all other $j \in C'$ such that $c(j) \neq$ NULL it is $\sigma(c(j),v) = \sigma(c(j),v')$.*

Example 2. With reference to the visualization context in Example 1, we consider three visualization types: pie chart, bubble chart, and heat map. The three suitability 7-tuples to be compared are shown in Table 3; the scores are excerpted from Table 2. Considering all the six specified coordinates (the interaction coordinate is not specified), it is bubble chart $\succ_C$ pie chart and heat map $\succ_C$ pie chart. Specifically, bubble chart flat-dominates pie chart because it is better on all coordinates except dependent type and cardinality, on which it is equivalent; similarly for heat map. On the other hand, there is no flat-dominance or flat-substitutability relationship between bubble chart and

Table 3. Suitability tuples for three visualization types with reference to the visualization context in Example 1

	pie chart	bubble chart	heat map
Goal: Comparison	neutral	fit	acceptable
Interaction: NULL	—	—	—
User: Tech	acceptable	fit	fit
Dimensionality: n-dim	unfit	fit	fit
Cardinality: High	discouraged	discouraged	fit
Independent Type: Interval	discouraged	fit	fit
Dependent Type: Ratio	fit	fit	fit

heat map because the first is better on the goal coordinate, while the second is better on the cardinality coordinate. So overall, if coordinate priorities are not considered, both bubble chart and heat map would belong to the skyline while pie chart would not. □

The last step is that of considering the priorities $\overset{c}{\succ}$ declared by the user as part of the visualization context. To this end we resort to the concept of *prioritized skyline* given in [13] and redefine dominance as follows.

Definition 4 (Dominance). *Given visualization context $c, \overset{c}{\succ}$ and two visualization types v and v', and given the set of coordinates $C' \subseteq C$, we say that v dominates v' on C' (denoted $v \vartriangleright_{C'} v'$) iff either (a) $v \succ_{max(C')} v'$ or (b) $(v \sim_{max(C')}) \wedge (v \vartriangleright_{C' \setminus max(C')} v')$, where $max(C')$ denotes the top coordinates in the $\overset{c}{\succ}$ order restricted to C'.*

Intuitively, if v is better that v' with reference to the coordinates that take highest priority for the user, then it is unconditionally better than v'; otherwise, if v is equivalent to v' with reference to those coordinates, we have to check if it is better with reference to the coordinates taking second priority, and so on.

Definition 5 (Skyline). *The skyline for $c, \overset{c}{\succ}$ is the set of visualization types in V that are not dominated by any other visualization type.*

Example 3. Considering again the visualization context in Examples 1 and 2, and taking now into account the coordinate priorities, it is bubble chart $\vartriangleright_C$ heat map $\vartriangleright_C$ pie chart. Indeed, since bubble chart and heat map are equivalent on the two top-priority coordinates (i.e., user and independent type), we have to check the second-priority coordinate (goal), on which bubble chart are better than heat map. So, taking into account priorities, the skyline only includes bubble chart. □

5   Evaluation and Conclusions

In this paper we have described an approach to automate the translation of the objectives declared by the user for visualizing the results of BDA into a set of most suitable visualization types. The approach enables users to specify a value for seven visualization coordinates, assigns a qualitative suitability score to each visualization type, then computes the skyline to determine the set of Pareto-optimal visualization types.

To evaluate our approach we have implemented a Java prototype whose interface supports the declaration of the visualization context and returns the prioritized skyline of visualization types. Then we have let the users of the three pilot applications of TOREADOR use this prototype to express a visualization context for their BDA use cases, and checked that they are satisfied with the visualization types proposed. For space reasons here we will describe only one use case out of the dozen use cases evaluated.

Fraud Detection. The goal of this use case is the identification of fraudulent clicks generated by bots in paid online advertising. Starting from a dataset describing the traffic through search engines and the related clickstreams, clustering and outlier detection algorithms are applied to determine a list of fraudulent IPs. The resulting data to be visualized describe the total number of clicks originated from the IPs of each country during 10 min slots of a single day. The visualization context declared by the users is

$$c(1) = \text{Trend}, \quad c(2) = \text{Filter}, \quad c(3) = \text{Lay},$$
$$c(4) = \text{n-dimensional}, \quad c(5) = \text{High}, \quad c(6) = \text{Ordinal}, \quad c(7) = \text{Ratio}$$
$$(1 \overset{c}{\sim} 3 \overset{c}{\sim} 4) \overset{c}{\succ} (2 \overset{c}{\sim} 5 \overset{c}{\sim} 6 \overset{c}{\sim} 7)$$

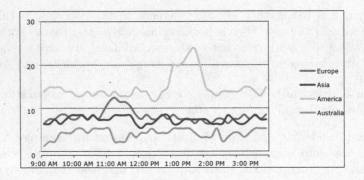

Fig. 3. Data visualization using a multiple line chart for the fraud detection use case

The skyline for the three top-priority coordinates of c includes multiple line chart, stacked line chart, and multiple line chart. However, when the remaining four coordinates are considered, only multiple line chart is left in the skyline (its suitability scores are neutral for filter, and fit for all other coordinate values). The resulting visualization is shown in Fig. 3, and was declared by the users to perfectly fit their needs.   □

Our future work mainly concerns the translation from the procedural to the deployment level of the TOREADOR platform. Specifically, one the user has chosen her preferred chart (e.g., bubble chart) among those suggested, and based on the types of the single (independent and dependent) data variables to be visualized, the system will support the user in mapping each data variable onto a specific dimension of the chart (e.g., first variable onto X axis, second variable onto Y axis, third variable onto bubble color, fourth variable onto bubble size).

References

1. Abela, A.: Advanced Presentations by Design. Pfeiffer, San Francisco (2008)
2. Ardagna, C., Bellandi, V., Damiani, E., Bezzi, M., Hebert, C.: A model-driven methodology for big data analytics-as-a-service. In: Proceedings of the IEEE International Congress on Big Data, Honolulu, Hawaii (2017)
3. Bertin, J.: Semiology of Graphics. Esri Press, Redlands (1983)
4. Börner, K.: Atlas of Knowledge: Anyone Can Map. MIT Press, Cambridge (2015)
5. Chandra, J., Madhu Shudan, S.: IBA graph selector algorithm for big data visualization using defense data set. Int. J. Sci. Eng. Res. 4(3), 1–7 (2013)
6. Dadzie, A.S., Rowe, M.: Approaches to visualising linked data: a survey. Semant. web 2(2), 89–124 (2011)
7. Few, S.: Show Me The Numbers: Designing Tables and Graphs to Enlighten. Analytics Press, Berkeley (2004)
8. Kano, N., Nobuhiku, S., Fumio, T., Shinichi, T.: Attractive quality and must-be quality. J. Jpn. Soc. Qual. Control 14(2), 39–48 (1984)
9. Keim, D.: Exploring big data using visual analytics. In: Proceedings of the EDBT/ICDT Workshops (2014)
10. Keim, D.A.: Information visualization and visual data mining. IEEE Trans. Vis. Comput. Graph. 8(1), 1–8 (2002)
11. Kleppe, A., Warmer, J., Bast, W.: MDA Explained - The Model Driven Architecture: Practice and Promise. Addison-Wesley, Boston (2003)
12. Marty, R.: Applied Security Visualization. Addison-Wesley, Boston (2009)
13. Mindolin, D., Chomicki, J.: Preference elicitation in prioritized skyline queries. VLDB J. 20(2), 157–182 (2011)
14. Peña, O., Aguilera, U., López-de-Ipiña, D.: Exploring LOD through metadata extraction and data-driven visualizations. Program 50(3), 270–287 (2016)
15. Russom, P.: Big data analytics. Technical report, TDWI Best Practices Report (2011)
16. Shneiderman, B.: The eyes have it: a task by data type taxonomy for information visualizations. In: Proceedings of the IEEE Symposium on Visual Languages, pp. 336–343 (1996)
17. Stevens, S.S.: On the theory of scales of measurement. Science 103(2684), 677–680 (1946)
18. Wehrend, S., Lewis, C.: A problem-oriented classification of visualization techniques. In: Proceedings of the IEEE Conference on Visualization, pp. 139–143 (1990)

A Four V's Design Approach of NoSQL Graph Databases

Jacky Akoka[1]([✉]), Isabelle Comyn-Wattiau[2], and Nicolas Prat[2]

[1] CEDRIC-CNAM & TEM-Institut Mines Telecom, Paris, France
jacky.akoka@lecnam.net
[2] ESSEC Business School, Cergy-Pontoise, France
{wattiau,prat}@essec.edu

Abstract. Big Data has been described as a four-dimensional model with Volume, Variety, Velocity, and Veracity. In this paper we discuss the potential of a model-driven approach (MDA) to tackle design issues of Big Data taking into account the effect of the four dimensions. Our approach considers NoSQL graph databases. The approach is applied to the case of Neo4j database. Our main contribution is an MDA methodology that enables to tackle the four V's dimensions described above. It consists of two major steps: (i) a forward engineering approach based on MDA as well as a set of transformations rules enabling the development of a conceptual, logical, and physical model for big data encompassing the four V's, (ii) a volume-guided approach supporting the generation of test bases dedicated to performance evaluation. We present an illustrative scenario of our forward engineering approach.

Keywords: Forward engineering · Big Data · NoSQL · Graph database · 4V's · Model-driven approach · Neo4j

1 Introduction

Social networks, sensors, mobile devices are generating data faster than ever. The Internet of Things is creating exponential growth in data. In addition, the speed of data generation increases at an exponential level. This gives rise to the concept of Big Data.

Big Data refers to large amount of data sets. In this paper we discuss the potential of a design approach taking into account the four Big Data characteristics, namely volume, variety, velocity, and veracity [1]. As stated by [2], "data is characterized not only by the enormous volume or the velocity of its generation but also by the heterogeneity, diversity and complexity". **Volume** refers to the quantity of data which can be larger than petabytes. **Variety** refers to the different types of data collected. It can be structured, semi-structured or unstructured. **Velocity** is the speed in which data is accessible. It is related to the rate of data generation and transmission. **Veracity** refers to the uncertainty of data due to factors such as data inconsistencies, or incompleteness [3]. [4] discuss the possibilities in creating trust in Big Data.

As a database grows in size and complexity and as it encompasses variety, velocity, and veracity issues, many traditional database systems suffer from serious performance issues. NoSQL databases have proven to be a better choice. However, most NoSQL

© Springer International Publishing AG 2017
S. de Cesare and U. Frank (Eds.): ER 2017 Workshops, LNCS 10651, pp. 58–68, 2017.
https://doi.org/10.1007/978-3-319-70625-2_6

databases are schemaless. They do not have any knowledge of the database schema, losing the benefits provided by these schemas. It is generally accepted that, when accessing the data, it is useful to know its schema. Moreover, query optimization and data integrity are best performed with schema-based databases. Therefore, data modelling can have an impact on performance, as well as on the ability to manage the four V's mentioned above. We argue that NoSQL databases need data models that ensure the proper storage and the relevant querying of the data, especially when we have to take into account the effect of volume, variety, velocity, and veracity.

This paper presents and illustrates a forward engineering approach enabling the design of NoSQL property graph databases incorporating the four V's at the early phase of the life-cycle. The approach is applied to the case of Neo4j database systems. It consists of two major steps: (i) a forward engineering approach based on MDA as well as a set of transformations rules enabling the development of a conceptual, logical, and physical models for big data encompassing the four V's, (ii) a volume-guided approach supporting the generation of test bases dedicated to performance evaluation. We present an illustrative scenario in order to assess the utility of our approach.

The rest of the paper is organized as follows. In Sect. 2 we present a state of the art on modeling NoSQL databases, especially for property graph databases. Section 3 describes our model-driven approach (MDA) dedicated to the design of NoSQL property graph databases including the four V's. It also describes the volume-guided approach supporting the generation of test bases enabling performance evaluation. The approaches including the transformation rules are illustrated in Sect. 4. Finally, Sect. 5 presents some conclusions as well as some perspectives of future research.

2   State of the Art

Big Data design issues have been the subject of several contributions. Several authors have proposed transformations from conceptual models or logical relational models into models of a specific NoSQL family or a specific NoSQL database management system (DBMS). Modeling NoSQL databases using forward engineering consists in going from a conceptual data model to a physical data model. [5] present an approach transforming relational model and RDF model to a property graph. [6] propose transformations that map a conceptual UML class diagram, with its associated OCL constraints, into a logical model of graph databases. [7] propose a database design methodology for NoSQL databases relying on NoAM models that can be implemented in several NoSQL databases. [8] propose an approach that allows to infer schemas from NoSQL databases. Regarding the use of MDA for Big Data, UML profiles have been proposed for conceptual, relational, and multidimensional models [9, 10]. As far as NoSQL DBMSs are concerned, several works have been proposed using UML. It is also one key ambition of the TOREADOR project [11].

Let us note that NoSQL DBMSs are generally schemaless, i.e. they do not support integrity constraints (ICs). The latter have been recognized as an effective method to express quality rules for databases. Common ICs in relational databases are defined on the schema. Several works study the usage of ICs in relational data warehouses [12]. Few works study ICs in the context of NoSQL databases. [13] define a mapping framework, with an associated query language, from relational to NoSQL DBMSs. As it can be seen from the contributions described above, NoSQL database design engineering is still in its infancy. To the best of our knowledge, there is no published approach on NoSQL property graph databases forward engineering explicitly taking into account the four dimensions (4V's) of Big Data. Similarly, there is no guided approach to assessing performance based on volume. This is precisely the purpose of our approach.

3 Our Approach

In this section, we aim to address the following research questions: (a) how to design conceptual, logical, and physical models dedicated to property graph databases incorporating the four V's; and (b) how to evaluate the performance of the graph database using the volume dimension. To answer these questions, we propose an MDA-based forward engineering approach enabling the development of a conceptual, logical, and physical model dedicated to NoSQL property graph databases incorporating the 4V's described above. We also present the transformation rules allowing us to move from one level to another. To achieve this result, we propose two meta-models, namely a conceptual and a logical graph meta-models. We also describe how, using the information regarding volume, we may generate automatically a test database to evaluate the performance of the graph database.

3.1   NoSQL Property Graph Databases

Graph databases model the database as a network structure containing nodes and edges representing relationship amongst nodes. The latter may also contain properties. Edges may also have their own properties. A relationship connects two nodes and may be directed. Relationships are identified by their names. They can be traversed in both directions. Comparing with the Entity-Relationship Model, a node corresponds to an entity, property of a node to an attribute, and edges to binary relationship between entities. Most available graph database systems, including Neo4j and OrientDB, have capability of storing semi-structured information.

3.2   The MDA Approach

To come up with the physical schemas, our approach is based on model-driven engineering. MDA facilitates modeling by differentiating between the modeling levels (conceptual, physical, and logical). It enables automated or semi-automated model transformations, and code generation. MDA has been applied successfully in several contexts [14, 15]. It distinguishes three levels: Computation Independent Model, Platform Independent Model, and Platform Specific Model. The Computation Independent Model corresponds to the user requirements which, generally, are expressed in natural language. The Platform Independent Model is the conceptual model. It can be declined into the different families of NoSQL databases: key-value, column family, graph, or document database. The Platform Specific Model is the implementation of the logical model in a specific database of the corresponding NoSQL family.

3.3 The Meta Models

In this paper, we propose a meta-model allowing us to represent an extended entity-relationship (EER) model. It supports n-ary relationships as well as ISA links between entities (Fig. 1). The information regarding the 4V's is another enrichment. The database designer has to anticipate the database size by assuming information on the number of occurrences that each entity and each relationship will contain, as well as the velocity of its growth. Variety is attached to attributes thanks to a detailed representation of data types. As an example, document is a specific data type. If such property is found in the conceptual model, a pure graph database like Neo4j will not be eligible but OrientDB may be the host of such data. Finally, veracity is represented both at the attribute level (characterizing trust of the designer in value quality) and at the aggregate level (characterizing the entities and the relationships) depending on the confidence of the database designer in sources of data.

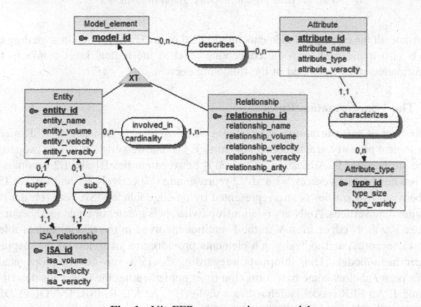

Fig. 1. V's EER conceptual meta-model

Our EER meta-model, supporting the representation of property graph databases, is inspired from [6]. The latter is enriched with specialization of ISA relationships and information regarding 4Vs. It allows the representation of the three basic elements of graphs, i.e. vertices, edges, and properties (Fig. 2). Edges connect tail vertices with head vertices. In order to support ISA relationships, we had to add an entity named ISA edge, subtype of the edge entity. EER conceptual model and property graph model are equally expressive. Thus, we were able to write down transformation rules allowing us

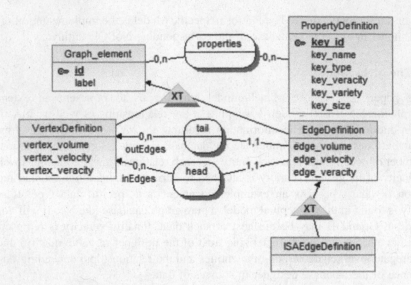

Fig. 2. EER logical property graph meta-model

to translate all the conceptual elements into logical ones. The information regarding the 4V's is also transferred from the conceptual level to the logical level thanks to the transformation rules described in the following section.

3.4   The Transformation Rules

The first set of transformation rules allows us to completely translate an EER model into a logical property graph model. An entity E becomes a vertex V whose properties are the attributes of E. A binary relationship R between entities E1 and E2 becomes an edge linking the two vertices V1 and V2 representing respectively E1 and E2. An ISA link between two entities is also represented by an edge labeled ISA between the two corresponding vertices. An N-ary relationship (when N is greater or equal to 3) becomes a vertex V with N edges from V to the N entities involved in the relationship. Table 1 crosses conceptual and logical graph elements providing an overview of the mapping between meta-models. Thus information regarding the 4V's may also be propagated.

We present below some transformation rules implementing the mappings described in Table 1. An EER model is defined as a tuple (EN, A, AT, R, ISA, INVOLVEDIN, DESCRIBES, CHARACTERIZES) where EN denotes the set of entities, A stands for the set of attributes, AT is the set of attribute types, R is the set of relationships, ISA is the set of ISA relationships, INVOLVEDIN stores the instances of all so-called relationships of our EER meta-model. In the same way, DESCRIBES and CHARACTERIZES are composed of couples linking respectively the related model elements. Adapted from [6] a graph database is defined as a tuple GD = (V; E; P), where V is the set of vertex definitions, E the set of edge definitions, and P the set of property definitions that compose the graph.

Table 1. Correspondence table between conceptual and logical graph model elements

MAPPING TABLE		VertexDefinition					EdgeDefinition					Tail	Head	PropertyDefinition					
		vertex id	vertex label	vertex volume	vertex velocity	vertex veracity	edge id	edge label	edge volume	edge velocity	edge veracity			key id	key name	key type	key veracity	key variety	key size
Entity	entity_id	X																	
	entity_name		X																
	entity_volume			X															
	entity_velocity				X														
	entity_veracity					X													
Relationship (when arity=2)	relationship_id						X												
	relationship_name							X											
	relationship_volume								X										
	relationship_velocity									X									
	relationship_veracity										X								
Relationship (when arity>2)	relationship_id						X												
	relationship_name		X					X											
	relationship_volume			X					X										
	relationship_velocity				X					X									
	relationship_veracity					X					X								
Attribute	attribute_id													X					
	attribute_name														X				
	attribute_type															X			
	attribute_veracity																X		
Attribute type	type_id													X					
	type_size																		X
	type_variety																	X	
Involved_in	isa_link											X	X						
Isa Relationship	isa_link							'ISA'											
	isa_volume								X										
	isa_velocity									X									
	isa_veracity										X								

As an example of rule, we provide below the transformation of an ISA relationship *i* between two entities *en1* and *en2*. It becomes an edge *e* labeled 'ISA' between *v1* and *v2* which are the nodes corresponding to *en1* and *en2*. The volume, velocity, and veracity of the ISA relationship *i* is transferred to edge *e*.

```
R1:  IF    ∃ i ∈ ISA AND ∃ en1, en2 ∈ EN | i=(en1,en2)
           AND ∃ v1 ∈ V | v1.id=en1.id AND ∃ v2 ∈ V | v2.id=en2.id
     THEN    let create e in E with e.label='ISA';
             TAIL:= TAIL ∪ {(e,v1)}; HEAD:= HEAD ∪ {(e,v2)};
             e.edge_volume:=i.isa_volume;
             e.edge_velocity:=i.isa_velocity;
             e.edge_veracity:=i.isa_veracity
```

The second rule example enables the transformation of N-ary relationships into vertices. The N-ary relationship *r* becomes a vertex *v*. In the same way, volume, velocity, and veracity information is propagated from *r* to *v*. Moreover, N edges are created to connect *v* to all N vertices *w* which represent all N entities *en* involved in *r*.

```
R2:  IF    ∃ r ∈ R | r.relationship_arity > 2
     THEN    let create v in V with v.label=r.relationship_name;
             v.vertex_volume:=r.relationship_volume;
             v.vertex_velocity:=r.relationship_velocity;
             v.vertex_veracity:=r.relationship_veracity;
          FOR EACH  (en,w) ∈ EN x V | (r,en) ∈ INVOLVEDIN  AND
w.id=en.id
             DO   let create e ∈ E | e.id= generateid();
                  TAIL:= TAIL ∪ {(e,v)}; HEAD:= HEAD ∪ {(e,w)};
          END FOR EACH
```

For space reasons, we cannot provide all the transformation rules. An additional output of our approach lies in its ability to predict the size and the main characteristics of the physical graph using information on volume. As an illustration, we may compute the conciseness of the graph and its connectivity [16]. The conciseness of the graph is computed using the following formula:

$$\sum_{v \in V} v.vertex\_volume + \sum_{e \in E} e.edge\_volume$$

In the same way, the connectivity of the graph is equal to:

$$\sum_{e \in E} e.edge\_volume \bigg/ \sum_{v \in V} v.vertex\_volume$$

As it will be shown in Sect. 4 (illustrative example), these metrics will constitute a mean to choose between different graph DBMS or between different storage options.

The second set of transformation rules allows us to transform a logical graph model into a script generating a Neo4j database. The script generates as many instances as defined by the *volume* value of the vertex or the edge. Executing the script generates the physical database. The resulting graph contains data with fictitious values. However the graph size is realistic facilitating performance tests.

Let us recall that the creation of a Neo4j graph using Cypher, which is Neo4j definition (and manipulation) language, is performed through one or several CREATE statements. As an example, the statement

`CREATE (shakespeare:Author {firstname:'William';lastname:'Shakespeare'})` generates the creation of a node whose label is Author. In addition, the statement `CREATE ((shakespeare)-[:BORN_IN]->(stratford))` generates an edge (label BORN_IN) between nodes `shakespeare` and `stratford`.

The first rule performs the generation of all vertices. For each vertex v belonging to V, it generates a number of vertices equal to *v.vertex_volume*. More precisely, the rule generates a Cypher statement which will be inserted in a script allowing to create the subsequent graph. The Cypher CREATE statement is obtained by concatenating all its constituents. The first FOR loop allows us to transform each vertex v. The second FOR loop uses i as an increment in order to generate exactly *vertex_volume* instances of the node v in the physical graph. The third FOR loop builds the string linking all the properties of v and uses j for generating instances of these properties. Thus nij is the value of the j-th property of the i-th instance of vertex v. In the same way, ni is the instance id of the i-th instance of vertex v.

```
FOR EACH v ∈ VertexDefinition
   FOR i:=1 to v.vertex_volume
      j:=1; statement:=' ';
      FOR EACH a ∈ v.properties  DO
            statement:=concatenation(statement, 'a.key:nij;');
            j:=j+1
         END FOR EACH
      statement:=concatenation('CREATE(ni: ',v.label,' {',statement,'})')
   END FOR
END FOR
```

The second rule performs the generation of all edges. For each edge *e* belonging to E, it generates a number of edges equal to *e.edge_volume*. More precisely, the rule generates a Cypher statement to be inserted in a script allowing to create the subsequent graph. The Cypher CREATE statement is obtained by concatenating all its constituents.

```
FOR EACH e ∈ EdgeDefinition
   FOR i:=1 to e.edge_volume
    j:=1; statement:=' ';
       FOR EACH a ∈ e.properties
          DO
             statement:=concatenation(statement,'a.key:eij;');
             j:=j+1
       END FOR EACH
       v1:=alea(e.tail);
       v2:=alea(e.head);
    statement:= concatenation('CREATE((',v1,')-[',e.label,'(',statement,
                ')]->(,v2,'))')
   END FOR
END FOR
```

The first FOR loop allows us to transform each edge *e*. The second FOR loop uses *i* as an increment in order to generate exactly *edge_volume* instances of the edge *e* in the physical graph. The third FOR loop builds the string linking all the properties of *e* and uses *j* for generating instances of these properties. Thus *eij* is the value of the *j*-th property of the *i*-th instance of edge *e*. We also have to connect this edge to two vertices belonging to the correct entities. To this end, we define the *alea* function which randomly extracts an instance of each entity respectively from *e.tail* and from *e.head*.

Putting together all the statements generated by the two rules results in a script that may then be executed leading to the creation of the physical Neo4j graph.

4 Illustrative Example

In order to illustrate our approach, we adapted the database example from [17] in order to better emphasize the salient aspects of our approach. This database stores information about scientific papers, their sources (journals, conferences), their authors, their affiliations, and the reflexive citation relationship between papers. Terms are keywords characterizing papers. We also added a reviewer entity linked to paper entity with a many-to-many relationship reviews. Author and reviewer are subtypes of researcher entity. We suppose that an author may have several affiliations but when he/she publishes a paper, he/she has to declare a unique affiliation. Thus, we represent a ternary relationship between papers, authors, and affiliations (Fig. 3). [17] also provides information on volumes that we used for testing our approach.

Applying the first set of transformation rules leads to the graph model at Fig. 4. The number of occurrences characterizes each node and each edge. We then generated a Neo4j database with as many instances as defined in the volume attribute of each vertex and each edge. Let us note that the edges are arbitrarily attached to vertices, respecting the labels and the volumes.

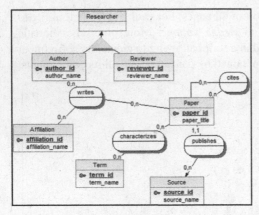

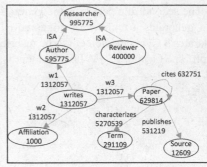

Fig. 4. Resulting graph logical model

Fig. 3. Example of conceptual model (adapted [17])

The conciseness of this graph is equal to 3342139 + 10733704 = 14075843. This parameter allows the database designer to estimate the size of the files and the possibility to use in-memory storage. Moreover, it allows him/her to choose between different physical organizations (e.g. flat text files as vertex-edge pairs, edge-vertex-vertex triples, or vertex-edge list records). Of course, this value may be computed even at the conceptual level since the number of graph instances directly results from the volume information available at the beginning of the design process. This is important since it allows the designer to anticipate the size of the graph database and to check if the target graph DBMS is eligible and if the storage capabilities are available.

In the same way, the connectivity of the graph is equal to 10733704/3342139 = 3.21. This score is higher than 1.5. It is considered to be good for processing [16].

5 Conclusions and Further Research

This paper presents an MDA approach dedicated to modeling issues of NoSQL property graph databases, encompassing the four V's, starting from a conceptual EER model. It is applied to Neo4j database. The approach consists mainly of two major steps. We perform an MDA-based forward engineering approach enabling the development of conceptual, logical, and physical models dedicated to NoSQL property graph databases incorporating the 4V's. We also present some transformation rules allowing us to move from one level to another. To achieve such result, we propose two meta-models, namely a conceptual and a logical meta-model. The second step, encompassing the generation of the physical database automatically generates a test database whose size is tuned using volume information. We present an illustrative example to assess the utility of our approach.

This work can be considered as part of a more comprehensive NoSQL roundtrip engineering process. It can be easily extended for roundtrip engineering of other NoSQL database systems incorporating the four V's. The following step will consist in propagating the information on velocity, variety, and veracity at the physical level of graph databases. Further work should embrace other types of NoSQL databases. Our approach will benefit from more experiments on large databases. To this end, we are currently developing a prototype implementing the meta-models and the rules.

References

1. Schroeck, M., Shockley, R., Smart, J., Romero-Morales, D., Tufano, P.: Analytics: the real-world use of big data. IBM Institute for Business Value - Executive Report (2012)
2. Llewellyn, A.: NASA Tournament Lab's Big Data Challenge, October 2012. https://open.nasa.gov/blog/2012/10/03/nasa-tournament-labs-big-data-challenge/
3. Lukoianova, T., Rubin, V.L.: Veracity roadmap: is big data objective, truthful and credible? In: Advances in Classification Research Online, vol. 24(1), pp. 4–15 (2014)
4. Sänger, J., et al.: Trust and big data: a roadmap for research. In: Database and Expert Systems Applications (DEXA), pp. 278–282, September 2014
5. Aggarwal, D., Davis, K.C.: Employing graph databases as a standardization model towards addressing heterogeneity. In: 2016 IEEE 17th International Conference on Information Reuse and Integration (IRI), pp. 198–207 (2016)
6. Daniel, G., Sunyé, G., Cabot, J.: UMLtoGraphDB: mapping conceptual schemas to graph databases. In: Comyn-Wattiau, I., Tanaka, K., Song, I.-Y., Yamamoto, S., Saeki, M. (eds.) ER 2016. LNCS, vol. 9974, pp. 430–444. Springer, Cham (2016). doi:10.1007/978-3-319-46397-1_33
7. Bugiotti, F., Cabibbo, L., Atzeni, P., Torlone, R.: Database design for NoSQL systems. In: Yu, E., Dobbie, G., Jarke, M., Purao, S. (eds.) ER 2014. LNCS, vol. 8824, pp. 223–231. Springer, Cham (2014). doi:10.1007/978-3-319-12206-9_18
8. Sevilla Ruiz, D., Morales, S.F., García Molina, J.: Inferring versioned schemas from NoSQL databases and its applications. In: Johannesson, P., Lee, M.L., Liddle, S.W., Opdahl, A.L., López, Ó.P. (eds.) ER 2015. LNCS, vol. 9381, pp. 467–480. Springer, Cham (2015). doi:10.1007/978-3-319-25264-3_35
9. Boulil, K., Bimonte, S., Pinet, F.: Conceptual model for spatial data cubes: a UML profile and its automatic implementation. Comput. Stand. Interfaces 38, 113–132 (2015)
10. Cuzzocrea, A., do N. Fidalgo, R.: Enhancing coverage and expressive power of spatial data warehousing modeling: the SDWM approach. In: Cuzzocrea, A., Dayal, U. (eds.) DaWaK 2012. LNCS, vol. 7448, pp. 15–29. Springer, Heidelberg (2012). doi:10.1007/978-3-642-32584-7_2
11. Toreador Project. http://www.toreador-project.eu/
12. Boulil, K., Bimonte, S., Pinet, F.: Spatial OLAP integrity constraints: from UML-based specification to automatic implementation: application to energetic data in agriculture. J. Dec. Syst. 23(4), 460–480 (2014)
13. Curé, O., Hecht, R., Le Duc, C., Lamolle, M.: Data integration over NoSQL stores using access path based mappings. In: Hameurlain, A., Liddle, Stephen W., Schewe, K.-D., Zhou, X. (eds.) DEXA 2011. LNCS, vol. 6860, pp. 481–495. Springer, Heidelberg (2011). doi:10.1007/978-3-642-23088-2_36
14. Abdelhedi, F., Brahim, A.A., Atigui, F., Zurfluh, G.: Big data and knowledge management: how to implement conceptual models in NoSQL systems? In: Proceedings of the 8th International Joint Conference on Knowledge Discovery, Knowledge Engineering and Knowledge Management, Porto, Portugal, pp. 235–240 (2016)

15. Prat, N., Akoka, J., Comyn-Wattiau, I.: An MDA approach to knowledge engineering. Expert Syst. Appl. **39**(12), 10420–10437 (2012)
16. Bouhali, R., Laurent, A.: Exploiting RDF open data using NoSQL graph databases. In: Chbeir, R., Manolopoulos, Y., Maglogiannis, I., Alhajj, R. (eds.) AIAI 2015. IAICT, vol. 458, pp. 177–190. Springer, Cham (2015). doi:10.1007/978-3-319-23868-5_13
17. Zhu, Y., Yan, E., Song, I.Y.: The use of a graph-based system to improve bibliographic information retrieval: system design, implementation, and evaluation. J. Assoc. Inf. Sci. Technol. **68**(2), 480–490 (2017)

Towards Efficient and Informative Omni-Channel Customer Relationship Management

Matthias Carnein[(✉)], Markus Heuchert, Leschek Homann, Heike Trautmann,
Gottfried Vossen, Jörg Becker, and Karsten Kraume

European Research Center for Information Systems (ERCIS),
University of Münster, Münster, Germany
{matthias.carnein,markus.heuchert,leschek.homann,heike.trautmann,
gottfried.vossen,joerg.becker,karsten.kraume}@ercis.uni-muenster.de

Abstract. Nowadays customers expect a seamless interaction with companies throughout all available communication channels. However, many companies rely on different software solutions to handle each channel, which leads to heterogeneous IT infrastructures and isolated data sources. Omni-Channel CRM is a holistic approach towards a unified view on the customer across all channels. This paper introduces three case studies which demonstrate challenges of omni-channel CRM and the value it can provide. The first case study shows how to integrate and visualise data from different sources which can support operational and strategic decision. In the second case study, a social media analysis approach is discussed which provides benefits by offering reports of service performance across channels. The third case study applies customer segmentation to an online fashion retailer in order to identify customer profiles.

Keywords: Omni-channel CRM · Big data analytics · Customer segmentation · Data architecture

1 Introduction

Creating a positive customer experience is of major importance in Customer Relationship Management (CRM) [1]. Building on the concept of customer-oriented [10] and service-centered [13] marketing, seamless interaction with companies throughout all communication channels builds a strong relationship between the brand and the customer. In order to consider customer needs and promote a positive customer experience, services need to be aligned appropriately. These developments are subsumed under the term omni-channel CRM, which does not only require multiple ways of contact (multi-channel CRM), but is characterized by an orchestration among the channels instead of an isolated management per channel [14]. Customers typically outpace companies in the adoption of new channels due to the strong momentum of private consumer electronics in form of smartphones or tablet technology. For this reason, companies

© Springer International Publishing AG 2017
S. de Cesare and U. Frank (Eds.): ER 2017 Workshops, LNCS 10651, pp. 69–78, 2017.
https://doi.org/10.1007/978-3-319-70625-2_7

struggle to extend their information systems and data management to maintain all channels consistently. Often, fast channel adoption from customers generates pressure and results in additional and poorly integrated systems. This paper aims to demonstrate the challenges and value that omni-channel integration can provide to customers and companies.

In a complex environment a holistic view on the customer is needed to improve the customer experience [9]. The concept of a *customer journey* models the relationship of a customer to a brand chronologically through a series of touch points. Each of these represents a part of the experience and can be a physical or digital contact with the brand that leaves a sentiment, i.e., a positive, negative or neutral experience for the customer. The experience can be based on rational or emotional impressions that may occur on purpose or by happenstance [3]. As every customer experience is different, personas are used to exemplify a typical customer journey [11].

While traditional channels such as voice are still dominant, a multitude of channels (e.g., social networks like Facebook, direct messengers like WhatsApp, or email) are demanded by customers. With this plethora of channels, customers tend to hop from one channel to another, while expecting no loss of information or service quality. The challenge is to identify the customer at each touch point. Figure 1 visualizes an example of a customer journey from the airline industry. Here, the customer checks the airline's website for booking information and addresses open questions in a Facebook messenger conversation. The subsequent online booking process is carried out on the website and additional luggage is added via email afterwards. At the airport, a delay of the airplane is communicated via WhatsApp. Lastly, the customer posts on the airline's Facebook wall about his experience. This exemplary journey highlights how customers can hop through channels while demanding equal service and no information loss regardless of channel. To demonstrate how a harmonious integration of channels can look like, we conduct and present three case studies in the omni-channel context.

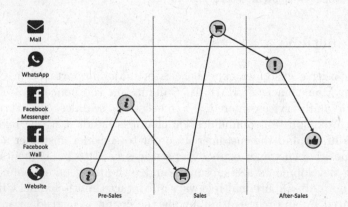

Fig. 1. Exemplary customer journey throughout a ticket purchase.

The remainder of this paper is structured as follows: Sect. 2 discusses a framework that structures omni-channel CRM from an analytical point of view. Based on this, three case studies are presented that demonstrate the challenges and value of omni-channel CRM. Section 3 presents a case study for data integration and Sect. 4 discusses how to visualise the integrated data. Then, Sect. 5 presents two case studies on analytical approaches for omni-channel scenarios to generate business value. Section 6 concludes with a summary of the results and an outlook.

2   Omni-Channel Framework

The utilization of customer data is crucial aspects of today's businesses and the interweaving of channel data can become a lever for success. The potential value in the data can be exploited through analytical approaches that enable an improved customer profiling, realise up- or cross-selling potential as well as personalisation of marketing efforts. Figure 2 depicts an extension of a framework [12] that structures the analysis and interplay of data, channels and the resulting benefits.

The process is considerably data-driven and makes use of different data sources and structures. Each data source has its own structure due to different storage requirements or characteristics. The different sources can for example be SQL, NoSQL or file systems and data can be either internally generated or provided by an external partner and updated in intervals or as a continuous data stream. The varying structure makes it necessary to transform the data between extraction and loading in order to bring it into compatible formats. Based on this data management, analytical approaches bridge the gap between

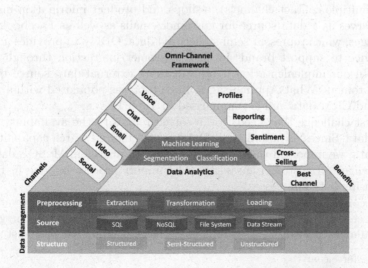

Fig. 2. Framework structuring the analysis of omni-channel data (cf. [12])

the channel or transactional data (left side) to enable benefits for the company or customer (right side). Concluding, the framework models all components in the analysis of omni-channel data and structures their relationships.

3 Data Management

In general data management comprises multiple areas such as data governance, data quality management, data architecture and design, database management and data warehousing [8]. In [12] we have introduced specific requirements to enable data management in omni-channel CRM. In particular, data persistency and unification of various data formats are relevant, as both play a key role for data integration. Data integration targets the issue most companies are confronted with nowadays: isolated data. This is based on the fact that a heterogeneous IT infrastructure across the various communication channels is common. For instance, a service centre might have chosen a particular software solution to handle voice and afterwards required an additional software to introduce the email channel. In the worst case, every channel is provided by a separate software solution. On top of that, software solutions can be provided as external services and therefore may require additional processing steps to integrate the data.

To demonstrate the challenges and benefits that this step poses, we have conducted a case study. In this academic project, we have emulated a real-life IT infrastructure which comprises three systems with different characteristics: Salesforce[1], Novomind[2] and OBI4wan[3]. We use this infrastructure as an exemplary case on how to integrate data from different sources. Salesforce provides a cloud-based CRM service and acts as a representative of a classical CRM system. Additionally, it manages the customers' data in a structured form within a relational database. Novomind offers products for service centres (i.e. call centres covering multiple contact channels), e-shops and product information management. It serves as a data source for customer emails as well as Facebook posts and messages, which represent semi-structured data. OBI4wan provides a cloud-based service to support brands in their customer interaction through online channels. In our implementation it represents an external data source to handle the customers' WhatsApp messages. Each tool was populated with artificial channel and CRM data and is summarised in Table 1.

The first challenge that needs to be solved relates to the availability of the retrieved data. Since Novomind and OBI4wan offer limited search capabilities for historic data, we imported the data into a separate relational MySQL database which manages all customer messages. The import process is done in regular intervals using batch processes.

The second challenge relates to the matching of customer information in order to create an integrated view on the data source. We base this integration on the available CRM data. Assuming that the phone number of a customer is

---

[1] www.salesforce.com.
[2] www.novomind.com.
[3] www.obi4wan.com.

Table 1. Data sources in the case study.

	Channel	System	Structure	Origin
CRM	–	Salesforce	Structured	Internal
Email	Email	Novomind	Semi-structured	Internal
Facebook	Social	Novomind	Semi-structured	External
WhatsApp	Social	OBI4wan	Semi-structured	External

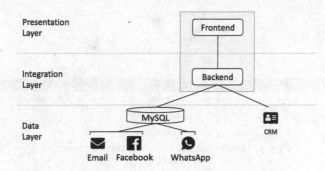

Fig. 3. Integration of different data sources in the case study.

available, we can easily use that to match the CRM data with WhatsApp chat messages. Similarly, if the email address is available, it can be used to match all emails, and the Facebook-ID can be used to match the social media account of a customer. In summary, we match all CRM attributes with the corresponding systems and combine them if available. An overview of our proposed architecture is provided in Fig. 3. Benefits of this approach are that each tool remains independent but its data is merged based on common attributes. In contrast to merging data sources permanently, this is a non-disruptive approach without the need to change working practices or the IT infrastructure. While this chapter has focused on the management of data, the following chapter targets its visualization.

4   Data Presentation

The integration of data across various platforms and channels can support strategic decisions made by management as well as the operational work. In a first step, the aggregated information can be visualised in order to provide a holistic view on the customer, its previous touch points as well as problems and purchase behaviour. At an operational level, this can improve customer service and help service agents by quickly providing them with important information about the customer. As a result, an agent can reply faster and thus reduce costs while providing a more tailored and relevant response.

To demonstrate the benefits of such an integrated view, we have developed a frontend which combines all the previously mentioned channels, while enriching

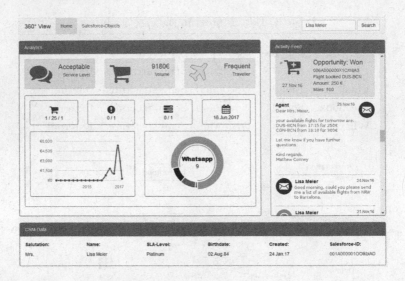

Fig. 4. Omni-channel view on the customer.

them with transactional and CRM data. The interface can support service agents in their daily work by showing them the relevant information about a customer at a glance. Figure 4 shows the of the proposed interface.

In its centre, a number of key characteristics of the customer are shown. Most importantly, its channel preferences are illustrated as an overview of past channel usage. This can be valuable information when reaching out to customers since it allows to choose channels that the customer prefers. Next to it, the value of previous purchases is shown which indicates the customer's importance for the company. In addition, a number of important statistics such as number of purchases, scheduled meetings, violations of the service level and travel frequency are shown.

On the right, an activity feed provides the most recent contact history with the customer. In other words, all Facebook wall posts, Facebook messages, WhatsApp messages and emails are shown. This component can provide a fast overview of previous touch points with the customer. It also facilitates the possibility to switch between channels without asking the customer the same questions over again. These features stress the value that only an integrated approach can provide. The interface is completed by CRM data at the bottom, i.e., by showing age, language or type of customer. It can be used stand-alone or embedded into existing systems.

5   Data Analytics

As shown above, the data from various channels can be used to provide a quick and complete overview of a customer. However, it also forms the basis of more sophisticated analyses. In the following, we present two case studies that demonstrate how to analyse data from different channels and to create benefits for customer and company.

5.1 Social Media Benchmarking of Customer Service

First, the collected channel data can be used for more detailed and efficient reporting. As an example, a core interest of a service centre operator is to keep the response times low. This is necessary in order to keep customers satisfied but often also mandatory due to contractual obligations. Collecting data about past and current cases allows easy comparison of such response times. This analysis can either focus one channel or span across multiple channels. In the following, we demonstrate insights gained from this approach by using the example of different social media platforms. In recent times, an increasing amount of service inquiries are made over the various social media platforms, most importantly Facebook and Twitter. Many customers use social media due to the immediacy, convenience and informal nature of the channel. To analyse this data we collected over 40 million service inquiries from Twitter and Facebook directed at the social media account of one of 250 companies. The collected data enables us to benchmark the service performance for a single channel or compare it to other channels. In addition, social media data allows easy benchmarking within and across industries.

As an example, Fig. 5 shows the average response time of four large airline carriers over the past months. In general, we observe that response times in social media are considerably faster than for traditional channels where customers might have to wait for several hours or days. Most messages in social media receive a response within the first four hours. When comparing data across the two social media platforms, it becomes obvious that Twitter messages receive a response considerably faster than Facebook messages. Often, the response time

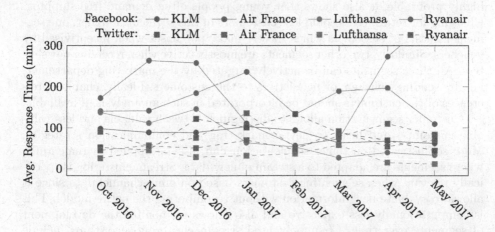

Fig. 5. Average response time of airlines in social media.

for Facebook posts is roughly 50% slower. Additionally, when comparing the four airlines, it becomes obvious that KLM and Lufthansa tend to respond much faster than Ryanair and Air France. This type of analysis can help to identify weaknesses in a company's social media strategy, highlight mismatches between the different channel strategies and compare to the performance of competitors. A detailed analysis of the collected data and customer service performance in social media is available in [5] where response times, response rates as well as conversation lengths are evaluated.

5.2 Customer Segmentation

Finally, channel and CRM data can also be used to reveal hidden customer segments. While the entire customer base can be viewed as heterogeneous, it is often possible to identify segments of homogeneous individuals that share similar behaviour or interests. The acknowledgement of customer segments allows to target each segment with specific products or marketing strategies. Since the customer's choice of channel is of particular interest in omni-channel CRM, this analysis can also incorporate channel usage in order to identify channel preferences and to refine the segments.

To demonstrate the benefits of this approach, we have performed customer segmentation for an online fashion retailer with more than 300, 000 customers. We have combined demographic and CRM data with transactional data and have used attributes such as the average order value, return rate, voucher usage and recent purchase frequency as features. To identify the segments we applied the well-known k-means [7] algorithm as well as Gaussian mixture models [15]. The analysis revealed nine different segments, each comprising between 5% and 20% of the customers. The characteristics of each segment can yield valuable insights into what preferences customers have and help to target them appropriately.

As an example, one segment contains young people that make expensive purchases but also extensively use the free-return option. While this segment is highly profitable, it also shows that young people often demand free shipping. Charging for return costs could lose customers of this segment. Further, one segment consists of customers which mostly make purchases when incentivised by vouchers. Similarly, two other segments are mostly active when receiving newsletters. All three segments can be actively targeted by the marketing department, i.e., by sending vouchers or newsletters to the customers. Ideally, channel preferences of the customers should be incorporated in such an analysis, if available.

Customer segmentation allows to distinguish between valuable and less valuable customers and helps to devise marketing and communication strategies. More sophisticated customer segmentation can also be applied incrementally, where segments are adapted to a stream of new data. Stream clustering is specifically relevant when segmenting customers based on omni-channel data since it allows to evaluate new information without recomputing the entire model. This is computationally less expensive and also allows to monitor the development of segments over time. Commonly used stream clustering algorithms include D-Stream [6] or DenStream [2]. An empirical comparison of different approaches is available in [4]. Stream clustering approaches usually incorporate an intermediate step, where the data stream is first summarised into a large number of preliminary clusters. This summary is then used with a traditional clustering algorithm to generate the final segments. A benefit of this approach is that relevant information is extracted from the stream without the need to retain every observation.

6   Conclusion and Outlook

In this paper we have introduced three case studies to highlight challenges and potential benefits of omni-channel CRM. All scenarios are structured using our comprehensive omni-channel framework. The first case study covers the data management layer of the framework to provide omni-channel insights. In particular, challenges for data integration are tackled by integrating data from different CRM system to create a unified view on the customer. This reduces heterogeneity of the IT landscape and can help to eliminate data silos. We have shown that this can support the service agent during his daily interaction with the customer and improve customer service. Once data silos are removed, new analytical approaches become applicable. To show this, we have introduced two case studies for analytical approaches in omni-channel CRM. The first case study demonstrates how public social media data can be utilized in order to compare the service performance and adjust the strategic alignment. On the one hand, it helps to identify weaknesses in the service performance across channels. On the other hand, it allows comparison among industries. The second case study demonstrates how to use CRM and transactional data for customer segmentation in order to identify groups of customers that share similar characteristics and behaviour. This allows to target each group individually by providing customised offerings while also increasing marketing efficiency for the company.

As our case study focuses on email and social media, potential future work could also integrate chat and voice data. Moreover, our data integration approach matches users based on available Facebook-ID, email and phone number. Here, more sophisticated approaches could be developed, e.g., automatically matching users by name or even picture. Finally, further analytical approaches could be employed such as automatically classifying user inquiries based on text analysis or customer segmentation based on streaming data.

References

1. Accenture: digital transformation in the age of the customer (2015). https://www.accenture.com/us-en/insight-digital-transformation-age-customer
2. Cao, F., Ester, M., Qian, W., Zhou, A.: Density-based clustering over an evolving data stream with noise. In: Conference on Data Mining (SIAM 2006), pp. 328–339 (2006)
3. Carbone, L.P., Haeckel, S.H.: Engineering customer experiences. Mark. Manag. **3**(3), 8–19 (1994)

4. Carnein, M., Assenmacher, D., Trautmann, H.: An empirical comparison of stream clustering algorithms. In: Proceedings of the ACM International Conference on Computing Frontiers (CF 2017), pp. 361–365 (2017)
5. Carnein, M., Homann, L., Trautmann, H., Vossen, G., Kraume, K.: Customer service in social media: an empirical study of the airline industry. In: Mitschang, B., Ritter, N., Schwarz, H., Klettke, M., Thor, A., Kopp, O., Wieland, M. (eds.) Proceedings of the 17th Conference on Database Systems for Business, Technology, and Web (BTW 2017), Stuttgart, Germany, pp. 33–40 (2017)
6. Chen, Y., Tu, L.: Density-based clustering for real-time stream data. In: Proceedings of the 13th ACM SIGKDD International Conference on Knowledge Discovery and Data Mining, KDD 2007, San Jose, California, USA, pp. 133–142. ACM (2007)
7. Hartigan, J.A., Wong, M.A.: A k-means clustering algorithm. Appl. Stat. **28**(1), 100–108 (1979)
8. DAMA International: The DAMA Guide to the Data Management Body of Knowledge - DAMA-DMBOK. Technics Publications, LLC, USA (2009)
9. Lemon, K.N., Verhoef, P.C.: Understanding customer experience throughout the customer journey. J. Mark. **80**(6), 69–96 (2016)
10. Peppers, D., Rogers, M.: The One to One Future. Currency, New York (1993)
11. Saffer, D.: Designing for Interaction: Creating Innovative Applications and Devices (Voices That Matter). New Riders, San Francisco (2009)
12. Trautmann, H., Vossen, G., Homann, L., Carnein, M., Kraume, K.: Challenges of data management and analytics in omni-channel CRM. Technical report 28, European Research Center for Information Systems, Münster, Germany (2017)
13. Vargo, S.L., Lusch, R.F.: Evolving to a new dominant logic for marketing. J. Mark. **68**(1), 1–17 (2004)
14. Verhoef, P.C., Kannan, P., Inman, J.J.: From multi-channel retailing to omni-channel retailing: introduction to the special issue on multi-channel retailing. J. Retail. **91**(2), 174–181 (2015)
15. Wedel, M., Kamakura, W.A.: Market Segmentation, 2nd edn. Springer, New York (2000). doi:10.1007/978-1-4615-4651-1

Stream Clustering of Chat Messages with Applications to Twitch Streams

Matthias Carnein[⊠], Dennis Assenmacher, and Heike Trautmann

University of Münster, Münster, Germany
{matthias.carnein,dennis.assenmacher,
heike.trautmann}@ercis.uni-muenster.de

Abstract. This paper proposes a new stream clustering algorithm for text streams. The algorithm combines concepts from stream clustering and text analysis in order to incrementally maintain a number of text droplets that represent topics within the stream. Our algorithm adapts to changes of topic over time and can handle noise and outliers gracefully by decaying the importance of irrelevant clusters. We demonstrate the performance of our approach by using more than one million real-world texts from the video streaming platform Twitch.tv.

Keywords: Data stream · Stream clustering · Text analysis · Text clustering · Twitch.tv

1 Introduction

Due to the increasing number of real-world applications producing data streams, the analysis of streaming data has become a key area of interest for research and practice. A core topic is the clustering of streaming data [9] which can be a valuable tool, e.g. for sensor analysis or customer segmentation [5]. However, a considerable amount of data nowadays comes in the form of text data such as e-mails, websites or chats. In order to analyse this data, it is necessary to build stream clustering algorithms that handle and analyse streams of texts [1].

In this paper, we combine ideas from text analysis and stream clustering in order to build an algorithm that can handle rapid data streams of text data. The texts can be of arbitrary length, language and content. Our approach uses a popular stream clustering approach [2] where the stream is first summarised, resulting in a number of small discussion threads. The algorithm adapts to changes in topics by forgetting outdated clusters while simultaneously identifying emerging clusters. Whenever necessary, the identified summaries can be *reclustered* by using a distance-based clustering algorithm to generate the overall topics. We demonstrate and evaluate our approach on real-world data streams by analysing more than one million chat messages from the video streaming platform Twitch.tv.

The remainder of this paper is structured as follows: Sect. 2 presents relevant approaches from text analysis and stream clustering. Next, Sect. 3 proposes

© Springer International Publishing AG 2017
S. de Cesare and U. Frank (Eds.): ER 2017 Workshops, LNCS 10651, pp. 79–88, 2017.
https://doi.org/10.1007/978-3-319-70625-2_8

our new algorithm which is able to cluster data streams of arbitrary text data. We evaluate the algorithm in Sect. 4 by clustering real-world chat messages. Finally, Sect. 5 concludes with a summary of the results and an outlook on future research.

2   Related Work

2.1   Text Analysis

A common approach when analysing text data is to vectorize the input texts and use the distance between the vectors as a measure of similarity. A simple approach is to use a vector of Term Frequencies (TFs) as the number of occurrences of a term t in a document d, i.e. $\mathrm{TF}(t, d) = |\{t \in d\}|$. However, a more popular approach is to count the number of occurrences for term-sequences in order to capture the context of words. Such term-sequences are called n-grams where n describes the size of the sequence.

Since not all terms provide the same amount of information, one can weight the TF with the Inverse Document Frequency (IDF). The IDF denotes whether a term is rare or common among all N available documents D. The underlying assumption is that rare terms provide more information. The IDF is often smoothed by taking the logarithm:

$$\mathrm{IDF}(t, D) = \log \left(\frac{N}{|\{d \in D : t \in d\}|} \right) \tag{1}$$

By multiplying both values, one obtains the so called Term Frequency – Inverse Document Frequency (TF-IDF) = $\mathrm{TF}(t, d) \cdot \mathrm{IDF}(t, D)$.

Finally, the distance between the resulting TF-IDF vectors $\boldsymbol{A}$ and $\boldsymbol{B}$ can then be computed as the cosine of the angle between them:

$$\cos(\theta) = \frac{\boldsymbol{A} \cdot \boldsymbol{B}}{\|\boldsymbol{A}\|\|\boldsymbol{B}\|}. \tag{2}$$

2.2   Stream Clustering

Due to the popularity of streaming applications, stream clustering has gained considerable attention in the past. While traditional cluster analysis assumes a fixed set of data, stream clustering works on a continuous and unbounded stream of new observations which cannot be permanently stored or ordered. Stream clustering aims to maintain a set of currently valid clusters, i.e. by removing outdated and learning emerging structures.

A recent example of a stream clustering algorithm is DBSTREAM [7]. In experimental results, the algorithm has shown the highest quality among popular stream clustering algorithms and fast computation time [5]. DBSTREAM employs a popular two-phase approach to cluster data streams [2]. Within an online phase, new data points are evaluated in real time and relevant summary statistics are captured. The result is a number of *micro-clusters* that summarise a large number of preliminary clusters in the data. During an offline phase, these micro-clusters can be 'reclustered' to derive the final clusters, called 'macro-clusters'.

In the online phase, DBSTREAM assigns a new observation to already existing micro-clusters if it lies within a radius threshold. When a micro-cluster absorbs the observation, its weight is increased and its centre is moved towards the new observation. If the observation lies within the distance threshold of multiple micro-clusters, it is added to all of them and a shared density between the clusters is stored as the number of points in the intersection of their radii. If the observation cannot be absorbed, it is used to initialize a new micro-cluster at its position. Since data streams may evolve over time, algorithms 'forget' outdated information [4]. DBSTREAM employs an exponential decay where the weights of micro-clusters are faded by $2^{-\lambda}$, in every time step. In fixed intervals, the algorithm evaluates the weight of all micro-clusters and removes those, whose weight decayed below a threshold. The offline component merges micro-clusters that have a high shared density to build the final clustering result.

2.3 Stream Text Clustering

One of the earliest algorithms to extend stream clustering to categorial and text data is ConStream (Condensation based Stream Clustering) [3]. The algorithm was proposed in 2006 and also follows the previously introduced two-phase clustering approach. Within the online phase, the algorithm maintains a set of summary statistics called cluster droplets which are similar to the concept of Clustering Features [10]. A cluster droplet is described as a tuple $(\overrightarrow{DF2}, \overrightarrow{DF1}, n, w(t), l)$ at a specific time t. $\overrightarrow{DF1}$ represents the sum of weighted occurrences for each word in a cluster. $\overrightarrow{DF2}$ stores, for each pair of terms, the sum of the weighted co-occurrences. This component is only used during the offline phase in order to analyse inter-cluster correlations. The remaining components of the tuple are scalar values that represent the number of observations within the cluster n, the total weight at a specific point in time $w(t)$ and a time stamp l that denotes the last time, the corresponding micro-cluster was updated.

Within the online phase, the similarity between a new data point $\overrightarrow{X}$ and all existing cluster droplets is calculated by using the cosine similarity on $\overrightarrow{DF1}$. If the similarity to the closest droplet is less than a predefined threshold, the algorithm searches for an inactive cluster which has not been updated in a while. It replaces the longest inactive cluster and initializes a new one for $\overrightarrow{X}$ instead. If no inactive cluster exists or the similarity is larger than the threshold, $\overrightarrow{X}$ is merged with the closest cluster. In the latter case, the updated droplet is faded by using the previously introduced decay function. In order to be able to compare and analyse different time horizons, a snapshot of all existing droplets is frequently persisted on secondary memory.

A review of relevant text stream algorithms can be found in [1]. However, we observed that within the popular research field of stream clustering, the aggregation of text data has received little attention and leaves room for improvement.

3   Efficient Stream Clustering of Text Data

In this section, we propose a new approach to cluster streams of text data. Our approach incrementally builds micro-clusters for text messages where each cluster maintains enough information to calculate a TF-IDF vector. The novelty our approach is as follows: (1) We combine state-of-art stream clustering concepts with text analysis using TF-IDF vectors, (2) we propose an incremental maintenance of the TF-IDF vectors, (3) we propose two fading approaches that efficiently maintain clusters. The approach is easy to implement and applicable to texts of arbitrary content, language and length.

The algorithm employs the widely accepted two-phase clustering approach where the online component updates the model whenever a new text is observed. The pseudo-code of this step is shown in Algorithm 1. We start by tokenizing the text and building the corresponding n-grams (line 2). Instead of using fixed size n-grams, we allow the user to specify a range between $n_{\min}$ and $n_{\max}$ and build all term-sequences within the given range. The result is an occurrence count for each n-gram in the text which can be considered its weight. Afterwards a temporary micro-cluster c is initialized for the new text. The micro-cluster is represented by the n-grams count, its weight is initialized to one and its time of last update is set to the current time t (line 3).

Algorithm 1. Update procedure

Require: t, $n_{\min}$, $n_{\max}$, message, λ, t_{gap}, r
1: **function** INSERT(message, $n_{\min}$, $n_{\max}$)
2:      ngrams $\leftarrow$ N-GRAMTOKENIZER(message, $n_{\min}$, $n_{\max}$)
3:      $c \leftarrow$ (ngrams, t, 1)   ▷ temporary micro-cluster
4:      **for** $i \leftarrow 1, ..., |MC|$ **do**
5:          $d_i \leftarrow$ COSINESIMILIARITY(mc_i, c)
6:      $j \leftarrow \arg\min_i(d_i)$ ▷ find closest micro-cluster
7:      **if** $d_j \leq r$ **then**
8:          $mc_j \leftarrow$ MERGE(c, mc_j, $2^{-\lambda\Delta t}$)
9:      **else**
10:          add c to MC
11:      **if** $t \bmod t_{gap} = 0$ **then**
12:          CLEANUP( )
13:      $t \leftarrow t + 1$

Next, we search for the closest existing micro-cluster to c by calculating the cosine similarity between the TF-IDF vector of c and all existing micro-clusters $mc_i \in MC$ (lines 4 – 6). While the TF-vector is directly available, the IDF vector can be calculated by summing the n-gram occurrences over all available micro-clusters.

If the closest micro-cluster is within a distance threshold r, we merge c into it (line 8). Micro-clusters can be easily merged by summing the number of occurrences per n-gram, summing the weight and setting the time of last update to the current time. If the closest micro-cluster is not within the distance-threshold, the temporary-cluster c is added as a new micro-cluster into the model (line 10).

In order to forget outdated clusters, we exponentially decay clusters using the fading function $2^{-\lambda \Delta t}$, where λ is a user chosen parameter and Δt is the time since the cluster was last updated. Whenever we update a cluster, e.g. by merging, we also update its weight. In addition, we also employ the same strategy to decay the frequency of each n-gram within the clusters.

Finally, we adjust the clustering every t_{gap} time-units to account for changes in weight and similarity of clusters (line 12). Algorithm 2 outlines its pseudo-code. First, we update the cluster-weight by applying the fading function to each micro-cluster (line 3). If a weight has decayed below $2^{-\lambda t_{gap}}$ we delete the micro-cluster (line 5). We choose this threshold, since it takes a new micro cluster at least t_{gap} time to decay to this weight [7]. Similarly, we update the weight of all n-grams within the micro-clusters by decaying their frequency and removing those that decayed below the same threshold (line 9). Finally, we evaluate the similarity between all pairs of micro-clusters. If clusters have moved into the distance threshold r, it is likely that they belong to the same cluster and we merge them as outlined above (line 10).

Algorithm 2. Cleanup procedure

1: **function** CLEANUP( )
Require: t_{gap}, MC, λ, r
  2:     **for each** $mc \in MC$ **do**   ▷ fade micro-cluster
  3:         WEIGHT(mc) ← WEIGHT(mc) $\cdot 2^{-\lambda \Delta t}$
  4:         **if** WEIGHT(mc) $\leq 2^{-\lambda t_{gap}}$ **then**
  5:             REMOVE(mc)
  6:         **for each** n-gram $\in mc$ **do**   ▷ fade n-grams
  7:             WEIGHT(n-gram) ← WEIGHT(n-gram) $\cdot 2^{-\lambda \Delta t}$
  8:             **if** WEIGHT(n-gram) $\leq 2^{-\lambda t_{gap}}$ **then**
  9:                 REMOVE(n-gram) from mc
 10:     Merge all mc_i, mc_j where COSINESIMILIARITY(mc_i, mc_j) $\leq r$

The result of the online component is a number of small micro-clusters, their weight and relevant n-grams that describe the cluster. This can be thought of as subtopics or small discussion threads in the stream. In order to derive the final topics from the micro-clusters we recluster them using a traditional clustering approach. To do so, we calculate a pair-wise distance matrix between all micro-clusters based on their cosine-similarity. We use the resulting matrix to apply hierarchical agglomerative clustering with complete linkage but any distance-based clustering approach is applicable.

4   Evaluation

4.1   Experimental Setup

In order to evaluate our approach, we implemented it as an extension to the stream package [6], a plugin for the statistical programming language R. In addition, we made use of real-world chat messages from the video streaming platform Twitch.tv. The platform is mostly hosting video-game related streams that regularly attract thousand of viewers which use a chat in order to communicate. Technologically, the chat is based on the Internet Relay Chat (IRC) protocol which allows real-time access to the chat.

For our analysis, we identified the ten most popular streams based on viewership and number of followers and collected the chat data from April, 12 to April, 14 2017. These channels produced more than one million text messages and on average we received 6.71 messages per second. Figure 1 indicates some seasonality in the data since streamers and viewers are less active around noon, as highlighted in red.

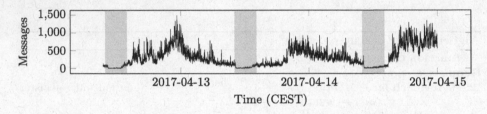

Fig. 1. Distribution of messages over the recorded period (Color figure online)

The goal of our analysis is to see whether we are able to identify reasonable discussion topics in the chat stream. Additionally, we try to find out whether the discussion contents of the streams differ, i.e. whether we are able to identify clusters which coincide with the different channels.

A core concept of our approach which differs from previous works is the capability to fade n-grams within a micro-cluster individually. This ensures that only relevant n-grams remain which greatly reduces the memory usage while maintaining comparable quality. In order to evaluate the performance of 'token fading', we executed our whole evaluation procedure on two different setups of the algorithm: the first setup explicitly utilizes the individual fading of tokens (cf. Algorithm 2 lines 6 – 9), whereas the second setup only fades the cluster-weights.

4.2   Evaluation Criteria

Evaluating cluster quality can be a challenging task since clustering aims at the discovery of unknown patterns. For this reason, quality is often evaluated using intrinsic information of the clusters, such as shape, size, and distance to other clusters. This approach is called internal evaluation and typically assumes that compact, spherical shapes indicate a higher quality. An example for this is the Silhouette Width as a measure of how similar an observations is to its own cluster, compared to other clusters. It is defined as $(b(i) - a(i))/\max\{a(i), b(i)\}$, where $a(i)$ is the average distance of i to other points in its cluster and $b(i)$ is the lowest average distance of i to points in another cluster. The silhouette width is typically averaged over all observations to derive a single index.

An additional approach to assess cluster quality is external evaluation. If a true partition of the data is known a priori, one can compare the clustering to the known groups. As an example, the purity of a cluster describes the proportion of points that belong to the majority class in the cluster. In our scenario it is difficult to apply external evaluation since the chat is not labelled by default. However, we can infer labels based on the channel or game that is played. As an example, we can label each message by the channel name to see whether we can evaluate whether topics differ across channels.

4.3 Tuning

In order to find appropriate parameter configurations for the proposed algorithm we apply iterated racing (`irace`) [8]. `irace` samples new parameter configurations and iteratively biases the sampling towards better solutions. We evaluate the performance of a configuration by using prequential evaluation [4,7] with a horizon of length $h = 100$. The idea is to evaluate the current model with the next h examples in the stream. Afterwards, the same examples are incorporated into the model before repeating the same process with the next h points. In our scenario, we use `irace` to find the configurations that yield the highest average purity over all horizons. We search $r \in [0,1]$, $\lambda \in [0,1]$, $t_{gap} \in \{50, ..., 1000\}$ and utilize unigrams to reduce the complexity. Since the purity can be arbitrarily improved by increasing the number of micro-clusters, we enforce an upper limit by discarding configurations that produce more than 500 micro-clusters. Due to the high computational complexity of this step, we restrict the parameter optimization to the first 100.000 observations in the stream. However, we observed that the results are stable for the remaining part of the stream. For our setup with token fading, the best solution was found for $r = 0.3593$, $\lambda = 0.7682$, $t_{gap} = 392$. Without token fading the best solution was $r = 0.1902$, $\lambda = 0.7105$, $t_{gap} = 329$.

4.4 Results

The results of our analysis shows that both setups of our algorithm produce very pure clusters (Fig. 2). For micro-clusters, the median purity of both setups is above 96% and both reach perfect purity at times while never dropping below 84%. When reclustering the results to macro clusters, the purity naturally drops but its median remains above 80%. This result shows that chat messages from the same channel are grouped into the same cluster and channels are rarely mixed.

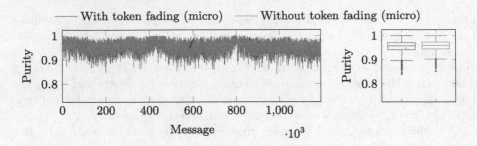

Fig. 2. Purity of clusters over time

Since the purity is very much influenced by the number of micro-clusters, we can observe that our parameter optimization yielded configuration that utilize close to 500 micro-clusters, i.e. our upper limit. Despite this strong compression of the original stream with 1.2 million messages, our algorithm retained enough information to group observations accurately.

When looking at the Average Silhouette Width (Fig. 3), we can observe a median width of around 0.3 indicating structures of medium quality. In peaks, the silhouette width can reach 0.8, indicating very strong structures but it can also drop to poor quality at times. As an example, we can observe severe decreases after 800.000 messages where the silhouette width decreases by almost 60%. This is likely explained by a swift change in topic, e.g. because channels stopped broadcasting.

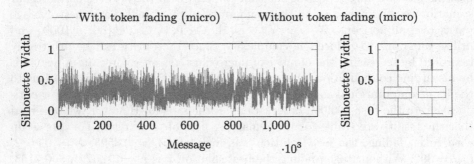

Fig. 3. Avg. silhouette width of clusters over time

All of the evaluation measures indicate that there is almost no difference in quality, when clustering was executed with or without n-gram fading. This observation supports the assumption that all the n-grams that were individually faded within a micro-cluster do not contribute to the clustering quality. In addition, our fading strategy requires less memory and reduces the median number of maintained n-grams by 78. However, we also observe that the λ parameter influences to what extent the TF vectors are reduced. For a high fading factor, the entire cluster will be removed before the effect of the n-gram fading can be observed.

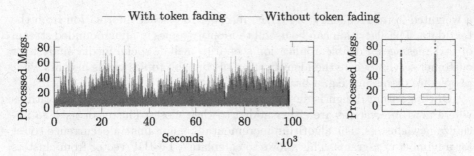

Fig. 4. Number of processed messages per second

Finally, we also evaluate number of messages that we were able to process every second on an Intel E5-2630 CPU with 2.2 Ghz (Fig. 4). On average, the algorithm was able to process 12 messages every second, however we observe many spikes in performance where almost 90 messages could be processed per second. Albeit fewer, we also observe the contrary, where performance slows down dramatically and only a single message is processed per second. The changes in processing speed can be explained by the varying complexity and length of chat messages. In order to evaluate whether our algorithm is able to process the data stream in real time, we compare the processing speed with the actual messages per second (Fig. 5). Even though there are times where the stream is faster than the processing speed, we observed that the average processing capability considerably surpasses the number of new messages in our data stream.

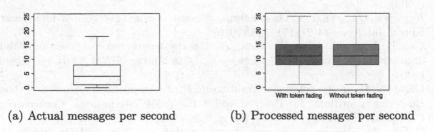

(a) Actual messages per second   (b) Processed messages per second

Fig. 5. Comparison of arrival and processing speed

5   Conclusion and Outlook

The analysis of text streams poses many challenges due to its rapid and unstructured nature. However, finding patterns and segments in such unstructured data streams can yield valuable insights and findings. In this paper we proposed a new stream clustering algorithm for text data. It utilizes the common concept of a weighted Term Frequency (TF-IDF) in order to extract information from the text data. The algorithm can be useful to identify topics in an unbounded stream of text messages. Possible application scenarios include social media analysis or customer service, where the identification of common topics yields insights what people are discussing, e.g. a brand.

Our algorithm design is similar to traditional stream clustering algorithms where new observations are either added to their closest cluster or used to initialize new clusters. The algorithm incrementally maintains an occurrence count for a number of n-grams. This allows to calculate a TF-IDF vector from clusters in order to calculate the similarity between them. Due to our design, this calculation comes with little computational overhead or memory consumption. In addition, clusters are exponentially decayed over time in order to forget outdated data.

Future work should compare our algorithm to alternative stream clustering approaches and evaluate our algorithm on longer texts, e.g. email data. In addition, other reclustering approaches than hierarchical clustering could be investigated. Finally, means to automatically and adaptively choose the required parameters should be pursued in order to make the application of the algorithm easier.

References

1. Aggarwal, C.C.: Mining text and social streams. ACM SIGKDD Explor. Newsl. **15**(2), 9–19 (2014)
2. Aggarwal, C.C., Han, J., Wang, J., Yu, P.S.: A framework for clustering evolving data streams. In: Proceedings of the 29th International Conference on Very Large Data Bases, VLDB 2003, vol. 29, pp. 81–92. VLDB Endowment, Berlin, Germany (2003)
3. Aggarwal, C.C., Yu, P.S.: On clustering massive text and categorical data streams. Knowl. Inf. Syst. **24**(2), 171–196 (2010)
4. Cao, F., Ester, M., Qian, W., Zhou, A.: Density-based clustering over an evolving data stream with noise. In: Conference on Data Mining (SIAM 2006), pp. 328–339 (2006)
5. Carnein, M., Assenmacher, D., Trautmann, H.: An empirical comparison of stream clustering algorithms. In: Proceedings of the ACM International Conference on Computing Frontiers (CF 2017), pp. 361–365 (2017)
6. Hahsler, M., Bolanos, M., Forrest, J.: stream: Infrastructure for Data Stream Mining (2015). https://cran.r-project.org/web/packages/stream/index.html
7. Hahsler, M., Bolaños, M.: Clustering data streams based on shared density between micro-clusters. IEEE Trans. Knowl. Data Eng. **28**(6), 1449–1461 (2016)
8. López-Ibáñez, M., Dubois-Lacoste, J., Pérez Cáceres, L., Stützle, T., Birattari, M.: The irace package: Iterated racing for automatic algorithm configuration. Oper. Res. Perspect. **3**, 43–58 (2016)
9. Silva, J.A., Faria, E.R., Barros, R.C., Hruschka, E.R., de Carvalho, A.C.O.L.F., Gama, J.: Data stream clustering: A survey. ACM Comput. Surv. **46**(1), 131–1331 (2013)
10. Zhang, T., Ramakrishnan, R., Livny, M.: BIRCH: An efficient data clustering databases method for very large. In: ACM SIGMOD International Conference on Management of Data, vol. 1, pp. 103–114 (1996)

MREBA 2017 - 4th International Workshop on Conceptual Modeling in Requirements and Business Analysis

Preface

It is our pleasure to welcome you to the fourth edition of the International Workshop on Conceptual Modeling in Requirements Engineering and Business Analysis (MREBA). MREBA 2017 occurred on November 6th, in the context of the International Conference on Conceptual Modeling (ER 2017), in the exciting city of Valencia, Spain.

The MREBA workshop aims to provide a forum for discussing the interplay between Requirements Engineering and Business Analysis topics and Conceptual Modeling. Requirements Engineering (RE) and Business Analysis (BA) are nowadays common practices within organizations, often applied in tandem. In this context, the use of Conceptual Modeling methods and languages are an essential practice, since both fields require conceptual models for analysis, reasoning, and documentation. In particular, the workshop focuses on how requirements modeling can be effectively used as part of Business Analysis and Systems Engineering.

This year, the workshop received 18 submissions, from which 8 papers have been accepted. The accepted papers are organized in two sections in this proceedings, reflecting the structure of the workshop's technical program. In a first section entitled *"Novel approaches to enterprises' and information systems' analysis"*, the topics of the four presented papers include: goal refinement and creativity in Goal-oriented Requirements Analysis, business process modeling assisted by natural language processing and enterprise architecture conceptual modeling domains. The second section is entitled *"Tools and techniques for formalizing strategy"* and gathers papers dealing with: ontologies and Semantic Web technologies for strategic decision making, and Key Performance Indicator (KPI) monitoring for goal-oriented Business Intelligence.

We deeply thank the authors of the submitted papers for their high-quality papers. We also thank the Program Committee members and additional reviewers for their effort and dedication in the review of the submitted works. And we finally thank the ER workshop chairs, PC chairs and the remaining of the organizing committee for their trust and support.

Renata Guizzardi
Eric-Oluf Svee
Jelena Zdravkovic
MREBA 2017 Organizers

Towards Consistent Demarcation of Enterprise Design Domains

Marné de Vries[(✉)] [iD]

University of Pretoria, Pretoria, South Africa
Marne.devries@up.ac.za

Abstract. This article supports the ideology that enterprise engineering (EE) could add more value if EE researchers focus on facilitating effective conversations within design teams to create a common understanding of the enterprise. One way of creating a common understanding is to define and demarcate enterprise design domains in a consistent way. Literature presents different conceptualisations for demarcating design domains, *without* using a systematic demarcation rationale. As an example, this article introduces Hoogervorst's approach and associated enterprise design domains to highlight practical difficulties when emerging *design principles* are applied to *four main design domains*, as defined by Hoogervorst. Based on the suggestion to apply the *basic system design process* to demarcate the main enterprise design domains in a *consistent way* and addressing the need for *additional* design domains, we present four alternative enterprise design domains, developed via *design science research*. We also demonstrate the usefulness of the new design domains by presenting several examples of enterprise design cycles that occur during enterprise design.

Keywords: Enterprise design · Design domains · Enterprise engineering

1 Introduction

Establishing design requirements is one of the most important elements in a design process and applied by many engineering disciplines [1, 2]. A more recent application of the design process, is to design the enterprise as an artefact, also termed enterprise engineering (EE) [3]. The enterprise engineer is mostly concerned with a holistic view of an enterprise [4] and the need to ensure enterprise-wide unity and integration [3, 5]. Yet, enterprise design is by no means simple, since enterprises rank amongst the highest in complexity, i.e. level eight on Boulding's [6] nine-level complexity scale. Even though there may be limits to formal enterprise design due to enterprise complexity, Hoogervorst [7] emphasises that the realisation of strategic intensions and successfully addressing areas of concern do not occur incidentally. Although most enterprises emerged in an ad hoc way, rather than by design [8], there is a need to govern enterprise evolution in a more systematic way [5]. Prior to governing its evolution, the enterprise design team needs to define those aspects or design domains that need to be governed. Yet, current enterprise design approaches vary in how they define different enterprise

© Springer International Publishing AG 2017
S. de Cesare and U. Frank (Eds.): ER 2017 Workshops, LNCS 10651, pp. 91–100, 2017.
https://doi.org/10.1007/978-3-319-70625-2_9

design domains/levels [4] and there is a lack of standardised terms, definitions, semantic rules and concepts to define the design domains [9].

This article explores the suggestion to use the *basic system design process* to demarcate four main enterprise design domains in a more *consistent* and *comprehensive* way. In addition, we *demonstrate the usefulness* of the newly-demarcated design domains by providing examples of several concurrent design cycles that occur during enterprise design.

The structure of the article is now discussed. Section 2 provides background on Hoogervorst's iterative enterprise design approach and associated enterprise design domains, as well as the *basic system design process* that is used as a means to demarcate enterprise design domains in a consistent way. Section 3 presents *design science research* as an appropriate research methodology for developing an *artefact*, namely a new *model* of constructs to represent enterprise design domains. Section 4 presents the *model* of constructs, as well as examples of concurrent design cycles that occur during enterprise design. Section 5 concludes with suggestions for future research.

2 Background

Section 2.1 provides background on the iterative enterprise design heuristic proposed by Hoogervorst [7]. In Sect. 2.2, we present theory on the *basic system design process* that may be useful when demarcating design domains in a *consistent* way.

2.1 Hoogervorst's Enterprise Design Heuristic and Practical Problems

Hoogervorst [5, 7] developed an approach that is iterative, emergent, creative and non-algorithmic. His approach contrasts with big-design-up-front (waterfall) approaches of the past and support the argument that stable requirements within a changing environment is an illusion [10], whereas domain knowledge of participating individuals is also emergent [11]. Hoogervorst's [12] approach supports Lapalme's [13] belief that EE will add more value if EE researchers focus on *effective conversations* within design teams, when they have a *common understanding of the enterprise* and emerging enterprise requirements. His multi-disciplinary inquisitive approach starts with the *strategic context*, defining preliminary *design aspects*, which are translated into *areas of concern* and *requirements*. Next architecture (called *design principles*) are defined per *design domain* by domain specialists to govern enterprise evolution [7]. Conceptualisation of appropriate *design domains* ensure that *design principles* can be applied to particular *design domains*, guiding the (re-)design of enterprise constructs within the particular *design domains* to operationalise key *areas of concern* [7]. Hoogervorst [5] presents *four main design domains*:

1. The *business domain* concerns the enterprise function, "having to do with topics such as products and services, customers and the interaction/relationship with them, the economic model underlying the business, and the relationships with the environment (sales channels, market, competitors, milieu, stakeholders)"[5, p. 299].

2. The *organisation domain* is part of enterprise *construction* and "concerns the internal arrangement of the enterprise, having for example to do with processes, employee behaviour, enterprise culture, management/leadership practices, and various structures and systems, such as regarding accounting, purchasing, payment, or employee evaluation" [5, p. 300]. Hoogervorst applies Dietz's [14] work to describe the *essence of the organisation domain* via aspect models. Aspect models are based on the transaction axiom, which states that the essence of enterprise operation consists of *human actor roles* that coordinate their actions around *production acts* to deliver goods and services to customers [14].
3. The *information domain* is also part of enterprise *construction* and consider aspects, such as "the structure and quality of information, the management of information (gathering, storage, distribution), and the utilisation of information [5, p. 300].
4. The *technology domain*, also part of enterprise *construction*, is "essential for business, organisational and informational support". Yet, Hoogervorst only highlights the need for *information technology* guidance [5, p. 300].

A previous study already experimented with Hoogervorst's approach, indicating difficulties during the process of sense-making, when emerging *design principles* had to be applied to Hoogervorst's demarcated *design domains* [15]. One of the reasons is that possible ambiguity exists between the *information domain* and the *organisation domain*. Hoogervorst's definition of the *information domain* includes constructs, such *production acts* (e.g. gathering, storage, distribution), that conceptually overlaps with the definition of the *organisation domain*. In addition, the study also highlighted that the existing *technology domain* focused on *information technology* alone, excluding other technologies that are also used to support the *organisation domain* [15]. As an example, a forklift enables/semi-automates the *production act* called *product relocation*. The study suggested that the main enterprise design domains be *redefined* in a more *consistent* way, i.e. based on the *basic system design process* [15].

2.2 The Basic System Design Process

The *basic system design process* is useful when an *object* system has to be designed within the context of a *using* system [14]. The design process has to start with knowledge about the *construction of the using system*, prior to eliciting functional requirements for a supporting *object system* [14]. Furthermore, the design process incorporates two main design activities, namely analysis (*determining requirements*) and synthesis (*devising specifications*) [16, 17]. Even though the *analysis* activities are distinguished from *synthesis* activities, Hoogervorst [7] emphasises that the activities are executed iteratively.

3   Research Method

This study applies *design science research* (DSR) as an appropriate research methodology to develop a *model* of constructs (see Fig. 1) to represent enterprise design domains that are based on a *consistent demarcation rationale*. In accordance with Gregor &

Hevner's [18] knowledge contribution framework, the model can be considered as an *improvement*, since a new solution (*model* of constructs for representing enterprise design domains) is developed for solving a known problem. Referring to the DSR steps of Peffers et al. [19], this article focuses on the first four steps of the DSR cycle as follows:

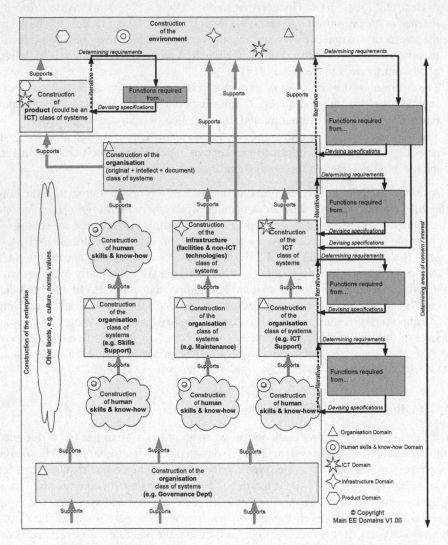

Fig. 1. Model of constructs to represent enterprise design domains

Identify a problem: Sect. 2 stated that current enterprise design approaches vary in how they define different enterprise design domains/levels [4] and there is a lack of standardised terms, definitions, semantic rules and concepts to define the design domains [9]. The Open Group [20] for instance defines four design domains (business, application, data and technology), whereas Hoogervorst [5] defines different design domains (business, organisation, information and technology). Also, a recent study that applied Hoogervorst's demarcation of design domains reported difficulties when emerging design principles had to be applied to Hoogervorst's demarcated design domains [15].

Define objectives of the solution: As a solution to the general class-of-problems regarding inconsistent demarcation of design domains and *lack of consistent demarcation rationale*, Sect. 2 suggests that the *basic system design process* is used as a means to demarcate the main enterprise design domains in a *consistent* and *comprehensive* way.

Design and develop: Based on the main concepts of the *basic system design process*, Sect. 4 presents a newly-developed artefact, namely a *model* of constructs to present enterprise design domains.

Demonstrate: Sect. 4 demonstrates the usefulness of the demarcated design domains by presenting several examples of concurrent design cycles performed during enterprise (re-)design.

Although this article only focuses on the *first four steps* of the DSR cycle, we elaborate on future work in Sect. 5.

4   Four Main Design Domains

In this section, we present the newly-demarcated design domains, using the basic system design process. Furthermore, we introduce different constructs of Fig. 1 and motivate their inclusion to represent four main enterprise design domains.

4.1   Constructs to Represent Design Domains

***Class of Systems* Constructs.**   Using the *basic system design process*, we identified an abstract *class of systems* (COSs) such that its *construction* is designed to support a using COSs. As an example, Dietz [16] identifies two enterprise COSs, the *ICT COSs* that is designed to support the *organisation COSs*. In addition, the *organisation COSs* should support the *environment COSs* or market context [16]. Other than the *organisation COSs* and *ICT COSs*, we added an *infrastructure COSs*, since Van der Meulen [15], reasoned that non-ICT technologies (e.g. facilities, utilities, machines and tooling), should also be designed to support the *organisation COSs*. Figure 1 illustrates *COSs* using light-grey-shaded rectangles and grey-shaded arrows labelled *Supports* to indicate supporting relationships between *COSs*. As indicated in Fig. 1, the *ICT COSs* may also directly *support* customers, business partners and suppliers within the environment [7]. Dietz [16] further partitions a COSs when he applies a layered nesting of systems, e.g. the *ICT COSs* may be further classified as two COSs: *software application COSs* supported by a *hardware COSs*. In addition, the *infrastructure COSs* may also incorporate subsystems, such as *material-handling COSs* and *energy-provisioning COSs* [21]. For simplicity reasons, Fig. 1 does not illustrate further partitioning.

Approach authors differ on whether *products* should be considered as an enterprise design domain. As an example, Bernard [22] and Williams [23] include the products domain, whereas The Open Group [20] excludes products as an enterprise design domain. Although not included as an enterprise construct, we acknowledge that a *product COSs* has a significant influence on enterprise design, since the *organisation COSs* should still support the construction of products that are sold to customers, as illustrated Fig. 1 via a *Supports* arrow. Figure 1 also indicates that the type of *product* sold to customers within the environment, may also be from the *ICT COSs*, since an enterprise may develop *software applications* as a *product* to their customers.

The *organisation COSs* at the bottom of Fig. 1, decoupled from other constructs to simplify the diagram, represents the *enterprise governance organisation* that needs to support the holistic and coherent design of all enterprise COSs and facets, but also need to be supported by *infrastructure, ICT* and *human skills & know-how*.

Facets Constructs. Hoogervorst [5] believes that some *facets*, such as culture, skills and learning requirements, may not be classifiable as *systems*, since a system is defined as elements that have influencing bonds within a particular composition [14, 24]. Yet, *facets* should still be designed as part of the enterprise and some do adhere to a life cycle, i.e. from identification, concept, requirement, design, implementation, operation and decommissioning [25]. Figure 1 depicts *facets* as light-grey-shaded cloud shapes. An example of a facet is *human skills & know-how*. An inductive analysis on existing enterprise design approaches [4] indicated that *human skills & know-how* should be designed and grown/developed intentionally.

Design Cycle Constructs. In Fig. 1, the dark-shaded rectangles with incoming and outgoing arrows represent the *basic system design process* that should be followed when designing an *object/provisioning* COSs/facet within the context of its *using* COSs/facet. Broken arrow-lines (labelled *Iterative* in Fig. 1) emphasise the iterative nature of analysis and design activities. The double-directed arrow-line on the right-hand side of Fig. 1 signifies the concurrent identification of *areas of concern/interest* that need to be addressed via enterprise design. Keeping the diagram simple, we did not illustrate all enterprise design cycles, but rather used grey-shaded arrows (labelled *Supports*) to indicate prominent relationships between constructs that would require iterative design cycles.

Design Domain Constructs. We already acknowledged different *COSs* and *facets* that need to be designed due to their support/provisioning relationships. Figure 1 also illustrates that COSs or facets have multiple supporting relationships. Although we duplicated some COSs to highlight different supportive relationships, small icons on Fig. 1 represents similar COSs/facets, i.e. *design domains*. The triangle icon represents the *organisation design domain*, the donut represents the *human skills & know-how design domain*, the 4-point star represents the *infrastructure design domain*, the 7-point star represents the *ICT design domain* and the hexagon represents the *product design domain*. As indicated before, the *product design domain* is currently not considered to be an *enterprise design domain*.

4.2 Domain-Related Design Cycles

Organisation Domain. Dietz [14] defines the *organisation COSs* as social systems, i.e. *actor roles*, implemented by human beings, form relationships due to their interactions and communications when they perform *production acts*. Dietz [14] suggests *aspect models* that represent the *essence of enterprise operation* in a coherent, comprehensive, consistent and concise way.

Following the *basic design process*, *organisation COSs* have to be designed within the context of its using COSs. According to Fig. 1, the *organisation COSs* shown as the top-most construct within the rectangle labelled *Construction of the enterprise,* can be designed within the context of different using COSs, i.e. (1) the *environment*, which encapsulates multiple COSs, and (2) *product COSs*. Figure 1 indicates that the *organisation COSs* also feature in support of other COSs, e.g. supporting the (1) construction of *human skills & know how*, (2) the *infrastructure COSs*, and (3) the *ICT COSs*.

When we consider, as an example, the design cycle that starts with the *construction of the environment* in Fig. 1, the enterprise design team considers the *construction of the environment*, i.e. *possible* products/services, other enterprises (markets, suppliers, partners, competitors, government institutions), citizens, channels, legislation, infrastructure, ICT and possible revenue to *determine requirements* and *functions* of the enterprise, which could be summarised in an *identify statement*. The identify statement provides meaning or sense-giving to enterprise stakeholders and may change over time [26]. An example of an identity statement for a department at a tertiary education institution is: *Encouraging blended learning, the engineering department offers tertiary education within the discipline of engineering, as well as quality research outputs.* Ostewalder [27] suggests that feasible *functions* of an enterprise can be specified in the form of a *business model canvass* in terms of a value proposition, key partners, key activities, key resources, customer relationships, customer segments, channels, cost structure and revenue streams. Based on the identified *functions*, the enterprise design team needs to *devise specifications* for *constructing the provisioning organisation COSs*, i.e. identifying and organising *actor roles* to perform appropriate *production acts*. During enterprise implementation, *production acts* and associated *actor roles* are usually grouped into departments, such as infrastructure, human resources, technology development, procurement, inbound logistics, operations, outbound logistics, marketing & sales and customer service [28].

Strategies for organising *production acts* may be influenced by several broad *areas of concern/interest* and even incidental concerns, which requires human sense-making and re-structuring to deal with the concerns [26]. Typical *areas of concern/interest* include profit, process excellence, customer orientation, and employee involvement.

ICT Domain. The ICT domain incorporates software applications, databases and ICT hardware [14]. Different representations are used to communicate ICT designs, such as *unified modelling language* (UML) models [29] and *wire-framing* models [30].

According to Fig. 1, an ICT system can be designed within the context of different using systems, i.e. (1) the construction of the *organisation COSs* and, (2) the construction of the *environment*.

When we consider, as an example, the design cycle that starts with the *construction of the organisation COSs* in Fig. 1, the enterprise design team considers the *construction of the organisation COSs*, i.e. the existing *actor roles* and their *production acts*, to *determine requirements* and *functions* of supporting *ICT COSs*. Based on the identified *functions*, the enterprise design team needs to *devise specifications* for the *construction of the provisioning ICT COSs*, i.e. providing *software applications, databases and hardware*.

Other than *functional concerns* to support the *construction of the using COSs*, typical *areas of concern/interest* for ICT COSs include interoperability, scalability, security and user friendliness.

Infrastructure Domain. Infrastructure entails facilities and other non-ICT technologies that support *actor roles* and their *production acts*. Enterprises within different industries may require different *representations* of infrastructure, based on the type of *production acts* that should be supported. The educational industry, for instance, may apply web-based *3D interactive campus models* to visualize learning facilities.

According to Fig. 1, the infrastructure COSs can be designed within the context of different using COSs, i.e. (1) the *construction of the organisation COSs* and (2) the *construction of the environment*.

When we consider, as an example, the design cycle that starts with the *construction of the organisation COSs* in Fig. 1, the enterprise design team considers the *construction of the organisation COSs*, i.e. the existing *production acts*, to *determine requirements* and *functions* of *provisioning infrastructure COSs*. Yet, according to Tompkins et al. [21], many other *organisation implementation decisions* also affect the provisioning infrastructure, such as packaging, service levels for spares, and delivery times. Based on the identified *functions*, the enterprise design team needs to *devise specifications* for the *construction of the provisioning infrastructure COSs*, i.e. providing *facilities, such as offices, factories and warehouses*.

Other than *functional concerns* to support the construction of the using COSs, typical *areas of concern/interest* for infrastructure include space utilisation, flexibility, upgradability, environmental friendliness, reliability, security, noise levels, vibrations, lighting, air quality and work space.

Human Skills & Know-how Domain. Human skills & know-how constitutes human *abilities and skills* required when executing *production acts*, as well as *coordination acts*. As indicated in Fig. 1, *human skills and know-how* cannot be defined as a *class of systems*, since there are no interactive parts, but need to be developed in support of the *organisation COSs*. Skills and know-how are often represented in *curricula vitae*.

When we consider, as an example, the first cycle that starts with the *construction of the organisation COSs* in Fig. 1, the enterprise design team considers the *construction of the organisation COSs*, i.e. the existing *production acts*, to *determine requirements* and *functions* of provisioning *human skills & know-how*. Based on the identified *functions*, the enterprise design team needs to *devise specifications* for required contextual knowledge, experience, skills and working styles (e.g. perseverance, stress resistance and self-control) to perform coordination acts and production acts.

Other than the *functional concerns* to support the construction of the using COSs, typical *areas of concern/interest* that should also be incorporated during design include *dynamic expansion of relevant knowledge*. Enterprises should not only encourage expansion of *skills and know-how* via formal training programmes, but also encourage facilitation of invisible learning environments for lifelong learning [31].

5   Conclusions and Future Research

Current enterprise design approaches vary in how they define different enterprise design domains/levels [4] and there is a lack of standardised terms, definitions, semantic rules and concepts to define the design domains [9]. Although the existing EE body of knowledge is mostly encapsulated in enterprise design approaches, many of the existing approaches do not provide a consistent demarcation rationale for their associated design domains, which impairs approach comparison [4].

Based on the ideology that EE could add more value if enterprise design teams create a common understanding of the enterprise [13, 32], the suggestion to apply the *basic system design process* to demarcate enterprise design domains in a *consistent way* and expanding the enterprise design scope by adding appropriate design domains [15], we developed a new *model* of constructs to represent enterprise design domains. In addition, we demonstrated the usefulness of the design domains by presenting several examples of enterprise design cycles that incorporate the design of object/provisioning COSs within the context of a using COSs.

Since this article only focused on the first four steps of the DSR cycle, future work is required to further demonstrate and evaluate the four main enterprise design domains. We suggest that a study, similar to [15], should be performed to apply the newly-demarcated design domains in association with the approach of Hoogervorst. Future research also needs to evaluate whether the newly-demarcated design domains are useful when combined with *other* enterprise design approaches.

References

1. Dym, C.L., Little, P.: Engineering Design, 3rd edn. Wiley, New York (2009)
2. Eggert, R.J.: Engineering design, 2nd edn. High Peak Press, Idaho (2010)
3. Dietz, J.L.G., Hoogervorst, J.A.P., Albani, A., Aveiro, D., Babkin, E., Barjis, J., Caetano, A., Huyments, P., Iijima, J., Van Kervel, S.J.H., Mulder, H., Op't Land, M., Proper, H.A., Sanz, J., Terlouw, L., Tribolet, J., Verelst, J., Winter, R.: The discipline of enterprise engineering. Int. J. Organ. Design Eng. 3(1), 86–114 (2013)
4. De Vries, M., Van der Merwe, A., Gerber, A.: Extending the enterprise evolution contextualisation model. Enterp. Inf. Syst. 11(6), 787–827 (2017)
5. Hoogervorst, J.A.P.: Enterprise Governance and Enterprise Engineering. Springer, Diemen (2009)
6. Boulding, K.E.: General systems theory: the skeleton of science. Manage. Sci. 2, 197–207 (1956)
7. Hoogervorst, J.A.P.: Applying Foundational Insights to Enterprise Governance and Enterprise Engineering. Springer, Berlin (to be published)

8. Giachetti, R.E.: Design of Enterprise Systems. CRC Press, Boca Raton (2010)
9. Von Rosing, M., Laurier, W.: An introduction to the business ontology. Int. J. Concept. Struct. Smart Appl. **3**(1), 20–41 (2015)
10. Proper, H.A., Lankhorst, M.M.: Enterprise architecture: towards essential sensemaking. Enterp. Model. Inf. Syst. Archit. **9**(1), 5–21 (2014)
11. Markus, M.L., Majchrzak, A., Gasser, L.: A design theory for systems that support emergent knowledge processes. MIS Q. **26**(3), 179–212 (2002)
12. Hoogervorst, J.A.P.: The imperative of employee-centric organising and its implications for enterprise engineering. J. Organ. Design Enterp. Eng. **1**, 1–16 (2017)
13. Lapalme, J.: Organizations (and organizing) are a technology that humans know very little about. Organ. Des. Enterp. Eng. **1**(1), 27–31 (2017)
14. Dietz, J.L.G.: Enterprise Ontology. Springer, Berlin (2006)
15. Van der Meulen, T.: Towards a useful DEMO-based enterprise engineering methodology, demonstrated at an agricultural enterprise. University of Pretoria (2017, unpublished)
16. Dietz, J.L.G., Albani, A.: Basic notions regarding business processes and supporting information systems. Requir. Eng. **10**, 175–183 (2005)
17. Dietz, J.L.G., Hoogervorst, J.A.P.: Enterprise ontology and enterprise architecture - how to let them evolve into effective complementary notions. GEAO J. Enterp. Archit. **1**, 1–19 (2007)
18. Gregor, S., Hevner, A.: Positioning and presenting design science research for maximum impact. MIS Q. **37**(2), 337–355 (2013)
19. Peffers, K., Tuunanen, T., Rothenberger, M., Chatterjee, S.: A design science research methodology for information systems research. J. MIS **24**(3), 45–77 (2008)
20. The Open Group: TOGAF 9.1. http://pubs.opengroup.org/architecture/togaf9-doc/arch/index.html. Accessed 17 June 2017
21. Tompkins, J.A., White, J.A., Bozer, Y.A., Tanchoco, J.M.A.: Facilities Planning, 4th edn. Wiley, USA (2010)
22. Bernard, S.A.: An Introduction to Enterprise Architecture EA3, 2nd edn. Authorhouse, Bloomington (2005)
23. Williams, T.J.: A Reference Model for Computer Integrated Manufacturing (CIM). Purdue Research Foundation, West Lafayette (1991)
24. Jackson, M.C.: Systems Thinking. Wiley, Chichester (2003)
25. Bernus, P., Nemes, L., Schmidt, G.: Handbook on Enterprise Architecture. Springer, Berlin (2003)
26. Weick, K.E., Sutcliffe, K.M., Obstfeld, D.: Organising and the process of sensemaking. Organ. Sci. **16**(4), 409–421 (2005)
27. Ostewalder, A., Pigneur, Y., Clark, T.: Business Model Generation, 1st edn. Wiley, Hoboken (2010)
28. Porter, M.E.: Competitive Advantage: Creating and Sustaining Superior Performance. The Free Press, New York (2004)
29. Theuerkorn, F.: Lightweight Enterprise Architectures. Auerbach Publications, New York (2005)
30. Garrett, J.J.: The Elements of User Experience: User-centered Design for the Web and Beyond, 2nd edn. New Riders Press, Berkeley (2011)
31. Marsick, V.J., Watkins, K.E.: Informal and indidental learning. New Dir. Adult Cont. Educ. **89**, 25–34 (2001)
32. Hoogervorst, J.A.P.: Foundational Insights for Enterprise Governance and Enterprise Engineering - Presenting the Employee-centric Theory of Organisation. Springer, Berlin (to be published)

Goal-Oriented Requirements Analysis Meets a Creativity Technique

Tomoo Kinoshita[✉], Shinpei Hayashi, and Motoshi Saeki

Department of Computer Science, Tokyo Institute of Technology,
Tokyo 152–8552, Japan
{kinoshita,hayashi,saeki}@se.cs.titech.ac.jp

Abstract. Goal-oriented requirements analysis (GORA) has been growing in the area of requirement engineering. It is one of the approaches
that elicits and analyzes stakeholders' requirements as goals to be
achieved, and develops an AND-OR graph, called a goal graph, as a
result of requirements elicitation. However, although it is important to
involve stakeholders' ideas and viewpoints during requirements elicitation, GORA still has a problem that their processes lack the deeper
participation of stakeholders. Regarding stakeholders' participation, creativity techniques have also become popular in requirements engineering. They aim to create novel and appropriate requirements by involving
stakeholders. One of these techniques, the KJ-method is a method which
organizes and associates novel ideas generated by Brainstorming. In this
paper, we present an approach to support stakeholders' participation
during GORA processes by transforming an affinity diagrams of the KJ-
method, into a goal graph, including transformation guidelines, and also
apply our approach to an example.

Keywords: Goal-Oriented Requirements Analysis · Creativity · KJ-
method

1 Introduction

Goal-oriented requirements analysis (GORA) has been developed regarding
requirements elicitation, analysis, and evolution in requirements engineering. It
assumes stakeholders' requirements as concepts of goals which require software
systems to achieve and are broken down into concrete means successively. By
modeling the stakeholders' requirements as goals, it aims to refine and complete
their requirements and make them easier to manage [4]. The resulting artifact
of GORA is an AND-OR graph, called a goal graph. Its nodes are goals and
edges between coarser-grained goals and finer-grained goals refined by them are
contribution links.

Various goal-oriented approaches have been developed, for example,
KAOS [3], i^* [15], and AGORA [1]. They are extensions of the general goal
model by associating goals with other elements, or giving several attributes to
goals. Thus, GORA has evolved with respect to requirements refinement.

© Springer International Publishing AG 2017
S. de Cesare and U. Frank (Eds.): ER 2017 Workshops, LNCS 10651, pp. 101–110, 2017.
https://doi.org/10.1007/978-3-319-70625-2_10

However, in goal-oriented approaches, there is still a problem that their processes do not include the support how stakeholders participate during requirements elicitation process. Generally, capturing stakeholders' ideas and viewpoints is important to elicit complete requirements which satisfy stakeholders' needs. However, often only requirements analysts have the role of gathering stakeholders' requirements and execute goal refinements because GORA is still too expert and difficult for stakeholders to understand. As a result, even if they could gather all information about stakeholders' requirements, they cannot involve stakeholders' ideas and viewpoints in the step of the goal refinement.

In terms of stakeholders participation, creativity techniques became more popular in requirements engineering. Sternberg's definition of creativity [11] is "the ability to produce work that is both novel and appropriate." Accordingly, creativity techniques in the requirements elicitation process are expected to produce novel and useful ideas for requirements [7]. Most creativity techniques are performed in a workshop form with stakeholders, which naturally encourages stakeholder participation.

Brainstorming [9] is the most commonly used creativity technique. It is a process where participants from different stakeholder groups attend to generate as many ideas as possible without focusing on any one in particular [16]. It goes on without any restrictions except keeping the following rules:

- Emphasize quantity,
- Do not criticize other people's ideas,
- Welcome wild ideas, and
- Combine and improve ideas.

The advantage of using Brainstorming is to make less complex to discover new and innovative solutions to existing problems. Brainstorming can generate by far the most ideas in creativity techniques [12].

Since Brainstorming expands ideas creatively, it is necessary to organize. Kawakita proposed the KJ-method to organize ideas created through Brainstorming and develop them more [2]. The KJ-method is performed in the following procedure:

1. The participants create ideas through Brainstorming. The first step is that the participants write down what has come to the mind on a card as many as possible. It allows being written down only one thing or idea for each card. According to Brainstorming, no criticism or no judgment on the importance should be made at this stage.
2. They group similar ideas and label the groups. The next step is to associate several ideas into one group. The participants group the created ideas by considering which ideas are semantically related to each other. After constructing a group, they discuss a label for grouped ideas which represents the essence of the group.
3. They find relationships among the groups and the ideas. The final step is to find some relationships between different groups or ideas. The participants discuss the existence of semantic relationships from each idea. The types of

relationships are causality, mutual causality, conflict, and constraint. Each type of relationships is represented by a different type of arrows between groups or ideas. The resulting artifact is called an affinity diagram.

However, since these creativity techniques focus on the creativeness of requirements, they didn't concern whether the elicited ideas cover from stakeholder's requirements to all the means for the system to achieve the requirements. In contrast, GORA can solve this problem by decomposing the ideas to means by composing them to requirements. Since creativity techniques and GORA can complement each other's problem, combining them can elicit more novel and complete requirements.

In this paper, we propose a novel method that connects a creativity technique and a goal-oriented approach while keeping the pure process of the creativity technique by connecting the resulting artifacts of them.

This paper consists of five sections. Section 2 explains our approach. In Sect. 3, we show an application example for our approach. Section 4 introduces related work, and Sect. 5 concludes this paper.

2   Process of Our Approach

2.1   Overview

We apply the KJ-method to the first step of the requirements elicitation and create an affinity diagram to transform into a goal graph because ideas generated by Brainstorming are too cluttered and difficult to form a goal graph directly. The KJ-method can solve the problem by creating affinity diagrams which can organize scattered ideas gathered by Brainstorming and have a group hierarchy which results in a clue for transforming into a goal graph.

As shown in Fig. 1, the process of our approach is summarized as follows:

1. Elicit ideas from stakeholders by Brainstorming,
2. According to the KJ-method, organize ideas and associate them (create an affinity diagram),
3. Transform the affinity diagram into a goal graph automatically, and
4. Modify the goal graph.

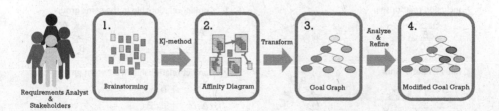

Fig. 1. Process of our approach.

The KJ-method naturally involves stakeholders because Brainstorming encourages association from others' ideas with stakeholders' insight without any restriction on their utterances. Also creating an affinity diagram does not hinder stakeholders participation because it needs to involve stakeholders to ask means of ideas to group or associate ideas.

2.2 Transformation Guidelines from an Affinity Diagram into a Goal Graph

This guidelines cover the transformation from all the general relationships in the KJ-method: constraint relationships, inclusion relationships, causal relationships, mutual causal relationships, and conflict relationships. An inclusion relationship means a relationship between a group and ideas included by it.

Create Goals from Ideas and Groups. Firstly, we create goals whose descriptions correspond to each description of all ideas and groups in the affinity diagram. By integrating all the ideas and the groups, we can elicit all ideas of stakeholders without missing them.

Transform Relationships into Contributions. Next, we transform relationships between ideas and groups in the affinity diagram into contributions between corresponding goals.

Transformations about each relationship are as Fig. 2 shows:

1. **Constraint relationship**. It means that the target idea requires a condition to achieve the source idea. It is the same as the goal model contribution.
   We create a contribution relationship whose source and target goals correspond to the target and source ideas respectively, and specify the refinement of the source goal for an AND-refinement. In Item 1 in Fig. 2, corresponding

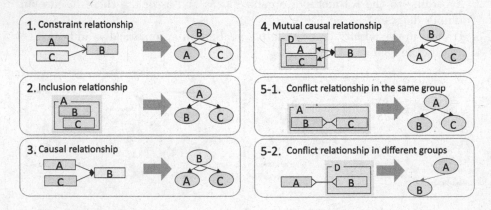

Fig. 2. Transformation guidelines for relationships to contributions.

goals to the ideas A and C are transformed into an AND-refinement of a corresponding goal of the idea B.

2. **Inclusion relationship**. The label of a group means the summary of ideas. This relationship is similar to refining goal granularity.

   We create a contribution relationship whose source and target goals are correspond to the group and the ideas in it respectively. Note that the refinement type of the source goal, i.e. an AND-refinement or an OR-refinement, is left undecided. In Item 2 in Fig. 2, corresponding goals to the ideas B and C are transformed to a refinement of a corresponding goal of the group A.

3. **Causal relationship**. It means that the source idea is a cause and the target idea is an effect. If the source idea is satisfied, the target idea is also satisfied. Hence, its meaning is the same as a contribution relationship in a goal model, i.e. the achievement of a source by a target.

   If the ideas in a causal relationship are in the same group, we create a contribution relationship whose source and target goals correspond to the target and source idea respectively, and specify the refinement of source goal for an AND-refinement. In Item 3 in Fig. 2, corresponding goals to the ideas A and C are transformed to an AND-refinement of a corresponding goal of the idea B.

4. **Mutual causal relationship**. A mutual causal relationship means that both of the ideas can be the cause and the effect. Therefore, this transformation is the same as a case of a causal relationship.

   We create a contribution relationship whose source and target goals correspond to the upper-level and lower-level ideas in group hierarchy in the mutual causal relationship respectively, and specify the refinement of source goal for an AND-refinement. In Item 4 in Fig. 2, corresponding goals to the ideas A and C are transformed to an AND-refinement of a corresponding goal of the idea B because the ideas A and C are lower in the group hierarchy rather than the idea B.

5. **Conflict relationship**. The goal model also has conflict links, but conflict ideas in the same group mean an exclusive-or situation. Consequently, a conflict relationship in the same group means an OR-refinement.

   If ideas in a conflict relationship are in the same group, we set the refinement of the goal which corresponds to the group to an OR-refinement. Otherwise, we create a conflict link whose source and target goals are correspond to upper-level and lower-level ideas in the group hierarchy in the conflict relationship respectively. In Item 5–1 in Fig. 2, corresponding goals to the ideas B and C are transformed to an OR-refinement of a corresponding goal of the group A because the ideas B and C are in the same group A. And also in Item 5–2 in Fig. 2, our guideline creates a conflict relationship between corresponding goals to the ideas A and B whose source goal is the idea A, because the idea B is lower in group hierarchy than the idea A.

We introduce only guidelines. Since the KJ-method includes Brainstorming, its activities are not restricted at all to create ideas. Needless to say, it is possible that transformations do not conform to the guidelines. Also, descriptions of ideas

in an affinity diagram can be inappropriate as goal descriptions in the goal graph
in most cases. Therefore, we need to modify such mismatch between the affinity
diagram and the goal graph.

3   Application Example

We conducted an application example to test our approach. We prepared three
stakeholders having different roles to perform the KJ-method in a face-to-face
discussion style. After that, we transformed the created affinity diagram into a
goal graph automatically according to our guidelines. Finally, we prepared two
new requirements analysts to refine the goal graph separately and discuss in the
differences together to reach consensus.

The affinity diagram created by the KJ-method in this example is shown in
Fig. 3, and the goal graph that was fixed by the analysts after transformation
is shown in Fig. 4. We developed and used a supporting tool to create affinity
diagrams and transform them into goal graphs automatically as an extension of
the goal graph editor for AGORA [10]. Each color of nodes means who created
the ideas in the affinity diagram and also corresponding goals in the goal graph.

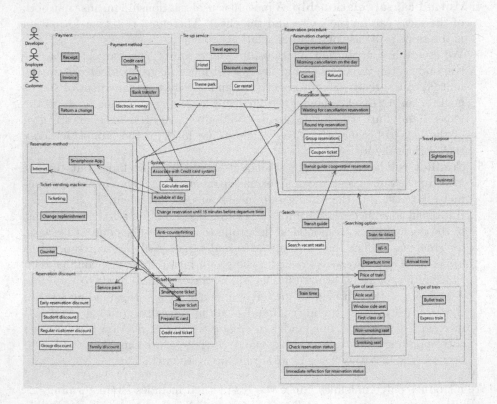

Fig. 3. Created affinity diagram in our application example.

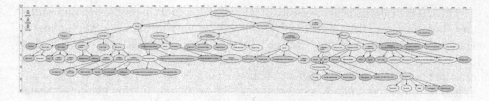

Fig. 4. Created goal graph in our application example.

Fig. 5. Added goals to align the granularity of the goal.

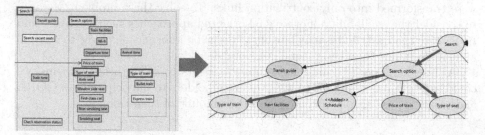

Fig. 6. Transformation example about an inclusion relationship.

We can see that all stakeholders were able to create ideas evenly because of the dispersion of colors.

A topic to elicit in this example was a train seat reservation system. The stakeholders were graduate students, and standpoints of them are a developer, a user, and an employee of the railroad company. Requirements analysts were also graduate students majored in requirements engineering who were familiar with GORA.

In this case, 76 goals were generated from the ideas in the affinity diagram on our tool. The requirements analysts judged that four of them were not necessary in the system. They needed to add 13 goals to align the granularity of goals because there were mixed grain ideas in the affinity diagram caused by the absence of restriction on the ideas. For example, as shown in Fig. 5, the granularity of the sub goal of the goal "Type of Seat" was judged inappropriate, so the analysts added the goals such as "Railroad coach" and so on.

They also discussed the necessity of contribution links according to our guide-lines, but most of them remained. For example, as Fig. 6 shows, the inclusion relationship from the group "Search" to the group "Search option" and that from "Search option" to "Type of train" and "Type of seat" were transformed into the contribution links without changes.

There appeared some problems about our approach in this example. Some participants pointed out that the horizontal width of the goal graph became very wide. It caused much human effort in understanding and modifying the goal graph. Since Brainstorming and the KJ-method do not restrict the topic, they can extend the idea side by side ceaselessly if it is related to the initial topic. As a countermeasure against this problem, it is possible to reduce the trouble of modifying the graph by decomposing the initial topic into several partial problems and integrating the results of the performance of our processes on each of them.

Another problem was that some descriptions of ideas are inappropriate as goals. Ideas created by Brainstorming are often written in the form of one word expressing just tasks, resources, and features. In addition, there was a problem that the semantic information about relationships were omitted because they were transformed into only contribution links. To solve these problem, we need to consider the availability of other goal-oriented approaches such as KAOS and i^* because their attributes may be able to describe information closer to original affinity diagrams than the general goal model.

Although we need to make some improvements to solve these problems, we could say that our approach enables the stakeholders to participate into GORA and create a goal graph with no gap in the amount of their opinion.

4   Related Work

Maiden et al. [7] reported some researches on creativity techniques in require-ments engineering and showed how to use creativity techniques in requirements elicitation. They also mapped the processes of established requirements activities onto creativity techniques processes.

Svensson et al. [12] compared Brainstorming with other creativity techniques, Hall of Fame, Constraint Removal, and Idea Box. They showed that Brainstorm-ing generates the most ideas, and Hall of Fame generates the most creativities. In this approach, we used Brainstorming in the KJ-method but we should inves-tigate the effectiveness of integration of other creativity techniques and GORA.

Mahaux et al. investigated how collaborative creativity techniques are effec-tive in requirements engineering from the factors of a collaboration team and an individual [6].

Leonardi et al. proposed an approach to elicit rich information about the users and the context, and complete system description by combining human-centered design and GORA [5].

Requirements elicitation and analysis by the KJ-method were performed by Takeda et al. [13]. They showed that the KJ-method is understandable without special knowledge of the notation. Also, they introduced an affinity diagram editor and conducted a case study of requirements elicitation and analysis for a novice system.

Ohshiro *et al.* proposed the method which integrates an affinity diagram into a goal graph [8]. However their method had many problems. First, the incompleteness of transformation about relationships in the affinity diagram. Last, their processes are all hand-powered, also, their method needs stakeholders to perform GORA for each goal by the KJ-method, which causes irrelevant refinement. We propose that requirements analysts conduct GORA for the created goal graph with stakeholders that are available to elicit complete requirements.

Wanderley *et al.* proposed a method to transform from mind maps into KAOS goal models for the same purpose as ours, stakeholders' participation in GORA [14]. The mind map method is certainly used as stakeholders' participation in requirements' elicitation. However, the associations of mind maps are too free to structure a goal model. In that respect, the group hierarchy and relationships of the KJ-method are more structured, and we can make a transformation close to the hierarchy. Also, although their method transformed into the KAOS goal model, our method is different in that it transforms into the general goal model.

5   Conclusion and Future Work

In our approach, requirements elicitation by the KJ-method and transformation from the created affinity diagram into a goal graph realizes stakeholders' participation in GORA. We also proposed transformation guidelines. We applied our approach to an example and found some problems in our approach such as the horizontal width of the goal graph, inappropriate goal descriptions, and omission of the information about relationships in affinity diagrams by transforming into single type links.

For future work, we try to solve such problems by decomposing the initial topic and integrating the result models and use of other goal-oriented methods such as KAOS or i^*. Also we need to evaluate the correctiveness of our guidelines and the difference from the conventional interview. It is also necessary to verify whether the process of our approach needs repeats recursively in the process of requirements elicitation and refinement. Furthermore, since our approach assumed to be use in a discussion style, it is useful to add the functions of speech recognition for the KJ-method and goal refinements.

Acknowledgements. This work was partly supported by JSPS Grants-in-Aid for Scientific Research Number JP15K00088.

References

1. Kaiya, H., Horai, H., Saeki, M.: AGORA: attributed goal-oriented requirements analysis method. In: Proceedings of the IEEE Joint International Conference on Requirements Engineering, pp. 13–22 (2002)

2. Kawakita, J.: KJ Method. Chuokoron-sha, Tokyo (1986)
3. van Lamsweerde, A.: Goal-oriented requirements engineering: a guided tour. In: Proceedings of the Fifth IEEE International Symposium on Requirements Engineering, pp. 249–262 (2001)
4. van Lamsweerde, A.: Requirements Engineering: From System Goals to UML Models to Software Specifications, 1st edn. Wiley Publishing, New York (2009)
5. Leonardi, C., Sabatucci, L., Susi, A., Zancanaro, M.: Design as intercultural dialogue: coupling human-centered design with requirement engineering methods. In: Campos, P., Graham, N., Jorge, J., Nunes, N., Palanque, P., Winckler, M. (eds.) INTERACT 2011. LNCS, vol. 6948, pp. 485–502. Springer, Heidelberg (2011). https://doi.org/10.1007/978-3-642-23765-2_34
6. Mahaux, M., Nguyen, L., Gotel, O., Mich, L., Mavin, A., Schmid, K.: Collaborative creativity in requirements engineering: analysis and practical advice. In: Proceedings of the IEEE 7th International Conference on Research Challenges in Information Science, pp. 1–10 (2013)
7. Maiden, N., Jones, S., Karlsen, K., Neill, R., Zachos, K., Milne, A.: Requirements engineering as creative problem solving: a research agenda for idea finding. In: Proceedings of the 18th IEEE International Requirements Engineering Conference, pp. 57–66 (2010)
8. Ohshiro, K., Watahiki, K., Saeki, M.: Integrating an idea generation method into a goal-oriented analysis method for requirements elicitation. In: Proceedings of the 12th Asia-Pacific Software Engineering Conference, pp. 9–17 (2005)
9. Osborn, A.: Applied Imagination: Principles and Procedures of Creative Thinking. Charles Scribner's Sons, New York (1957)
10. Saeki, M., Hayashi, S., Kaiya, H.: A tool for attributed goal-oriented requirements analysis. In: Proceedings of the IEEE/ACM International Conference on Automated Software Engineering, pp. 674–676 (2009)
11. Sternberg, R.J.: Handbook of Creativity. Cambridge University Press, New York (1999)
12. Svensson, R.B., Taghavianfar, M.: Selecting creativity techniques for creative requirements: an evaluation of four techniques using creativity workshops. In: Proceedings of the IEEE 23rd International Requirements Engineering Conference, pp. 66–75 (2015)
13. Takeda, N., Shiomi, A., Kawai, K., Ohiwa, H.: Requirement analysis by the KJ editor. In: Proceedings of the IEEE International Symposium on Requirements Engineering, pp. 98–101 (1993)
14. Wanderley, F., Araujo, J.: Generating goal-oriented models from creative requirements using model driven engineering. In: Proceedings of the 3rd International Workshop on Model-Driven Requirements Engineering, pp. 1–9 (2013)
15. Yu, E.S.K.: Towards modelling and reasoning support for early-phase requirements engineering. In: Proceedings of the Third IEEE International Symposium on Requirements Engineering, pp. 226–235 (1997)
16. Zowghi, D., Coulin, C.: Requirements elicitation: a survey of techniques, approaches, and tools. In: Aurum, A., Wohlin, C. (eds.) Engineering and Managing Software Requirements, pp. 19–46. Springer, Heidelberg (2005)

Towards an Ontology for Strategic Decision Making: The Case of Quality in Rapid Software Development Projects

Cristina Gómez[1(✉)], Claudia Ayala[1], Xavier Franch[1], Lidia López[1], Woubshet Behutiye[2], and Silverio Martínez-Fernández[3]

[1] Universitat Politècnica de Catalunya (UPC), Barcelona, Spain
{cristina,cayala,franch,llopez}@essi.upc.edu
[2] University of Oulu, Oulu, Finland
Woubshet.Behutiye@oulu.fi
[3] Fraunhofer IESE, Kaiserslautern, Germany
Silverio.Martinez@iese.fraunhofer.de

Abstract. Strategic decision making is the process of selecting a logical and informed choice from the alternative options based on key strategic indicators determining the success of a specific organization strategy. To support this process and provide a common underlying language, in this work, we present an empirically-grounded ontology to support different strategic decision-making processes and extend the ontology to cover the context of managing quality in Rapid Software Development projects. We illustrate the complete ontology with an example.

Keywords: Rapid Software Development · Strategic decision-making · Ontology

1 Introduction

Decision making is the process of selecting a logical and informed choice from the available options. When the logical choice is based on key factors determining the success of a specific organization-oriented strategy, this process is called strategic decision making. Strategic decisions are important because without them actions will not be planned to follow organizations' strategies.

Nowadays, techniques for decision making are applied to several fields such as business management and software engineering among others. In addition to these techniques, there are tools to help decision makers in the process of making decisions. Examples of these tools are decision support systems to cope with decision-making activities, and strategic dashboards to provide a view of strategic indicators.

A way of providing a common underlying language integrating the concepts to manage decisions and to handle strategic indicators is to define an ontology. This ontology enables structuring the knowledge in a way that favors its understanding and communication and, consequently, it could be used as a basis for the construction of

© Springer International Publishing AG 2017
S. de Cesare and U. Frank (Eds.): ER 2017 Workshops, LNCS 10651, pp. 111–121, 2017.
https://doi.org/10.1007/978-3-319-70625-2_11

tools supporting strategic decision making. Moreover, this ontology will be used, in future versions, to provide reasoning capabilities such as suggestions on demand.

The goal of this paper is twofold: (1) to present a preliminary *Strategic Decision-Making* (SDM) ontology to support different strategic decision-making processes and (2) to extend the SDM ontology to cover the context of managing quality in Rapid Software Development (RSD) projects.

The SDM ontology identifies the key terms through a glossary and their relationships through a conceptual model, represented with a UML class diagram and associated integrity constraints. We illustrate the possible use of the SDM ontology for assuring quality in RSD within the context of the research and innovation European H2020 project Q-Rapids[1] and provide an example for this context.

The rest of the paper is structured as follows. Section 2 discusses the related work in the area and Sect. 3 explains the research approach followed to define the SDM ontology. Section 4 elaborates the generic SDM ontology. Section 5 extends the SDM ontology to cover the context of the Q-Rapids project and presents an example. Finally, Sect. 6 summarizes the conclusions and identifies future work.

2 Related Work

We reviewed the extensive literature on how to model and measure the strategy of an organization. Basically, we are interested in identifying the concepts of strategic decision making, which are relevant for decision makers, and mapping those concepts to the context of quality requirements in RSD. We used these concepts to model and measure strategic indicators for supporting decision making.

The Balance Scorecard (BSC), proposed by Kaplan et al. [1], is a business framework used for describing and measuring an organization's strategy and for tracking the actions taken to improve the results. In this sense, the BSC proposes to define strategic objectives and key performance indicators (KPIs) to measure those strategic objectives and actions that the organization should take into account to achieve the objectives. The BSC has been applied in the context of measuring software quality[2].

The Business Motivation Metamodel (BMM) [2] provides concepts for developing, communicating and managing business plans in an organized manner. Specifically, the BMM defines concepts such as strategy and goals. In the same sense, the Business Intelligence Model[3] (BIM) [3] provides constructs for modelling business organizations at strategic level. In particular, concepts such as actors and KPIs are defined.

KPI, as a way of monitoring, is a crucial concept that has received a lot of attention in the literature. There exist catalogues of KPIs to measure several aspects of an organization. For instance, the Scoro work management software solution[4] provides 16

---

[1] Q-Rapids' website: http://q-rapids.eu/.

[2] http://www.bscdesigner.com/bsc-for-software-quality-guide.htm.

[3] http://www.cs.toronto.edu/~jm/bim/.

[4] https://www.scoro.com/blog/16-essential-project-kpis/.

essential project KPIs to track a project's performance (e.g., return on investment, overdue project tasks/crossed deadlines). Enfocus Solutions[5] defines KPIs for business analysis and project management (e.g., project stakeholder satisfaction index, and number of milestones missed). Besides, CBS[6] (Center for Business Practices) enumerates a comprehensive list of measures of project management and value in the context of IT organizations (e.g., average time to repair a defect, alignment to strategic business goals). An ontology in which KPIs are described together with their mathematics formulas is explained in [4].

In [5], Maté et al. present a semi-automatic approach that performs a partial search guided by the KPIs of the company, generating queries required during the monitoring process to discover the existence of problems and where they are located. Another approach that provides decision makers with an integrated view of strategic business objectives linked to conceptual data warehouses KPIs is presented in [6].

Regarding the decision-making process, the Decision Model and Notation (DMN) metamodel [7] provides the constructs that are needed to model decisions. DMN defines concepts such as decision, decision maker and knowledge requirement.

Previous works provide meta-models, ontologies and constructs representing concepts as KPIs, decision, and actions among others. However, there is not an integrated ontology establishing relationships among those concepts, which is our purpose.

3 Research Approach

This work was carried out under the context of the Q-Rapids European project [8] that aims to improve the management of quality requirements in RSD processes. To achieve this goal, the project aims to promote a highly informative dashboard to support making strategic decisions in rapid cycles. Such a dashboard will be based on the extraction and analysis of information to systematically and continuously assess software quality using a set of quality-related indicators based on GQM + Strategies™ [9] in order to support decision-making processes. Information will be extracted from diverse repositories containing information about the software development process (e.g. bug tracking systems), runtime behavior of the software system (e.g. software monitors) and system usage (e.g. end user feedback).

In order to characterize the information to be managed by the intended dashboard, we devised the use of ontologies. The backbone ontology of the intended dashboard is the SDM ontology that can be generalized to other strategic decision-making processes. In addition, we have extended the ontology with two additional packages (see Fig. 1): Quality Assessment (QA) and Rapid Software Development Process (RSDP) to cover the particular domain of the Q-Rapids project that is supporting decision making related to quality requirements in RSD processes.

The conceptualization of the ontologies has followed an iterative and incremental process based on the Methontology conceptualization phase [10]. The proposed

---

[5] http://enfocussolutions.com/kpis-for-business-analysis-and-project-management/.

[6] http://www.pmsolutions.com/audio/PM_Performance_and_Value_List_of_Measures.pdf.

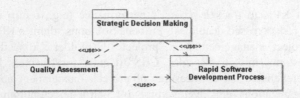

Fig. 1. UML package diagram for strategic decision making of quality in RSD

ontologies have been empirically-grounded on the basis of the study of four Q-Rapids industrial partners from different European regions and sizes. Q-Rapids industrial partners provide use cases to collect empirical data needed to solidify the objectives of the project and to serve as basis to implement a validation plan. To gather this data, we conducted semi-structured interviews at the industrial partner's premises, in situ observations and accessed to some of their repositories and tools to analyze the data that could be exploited. The main activities performed to define the ontologies were:

- *Activity 1: Definition of the ontology structure:* Identification of the ontology packages to group concepts with related semantics and to provide a namespace for the grouped elements. We decided to create three different packages (see Fig. 1) in order to address and link three relevant problems.
- *Activity 2: Extraction of terms relevant for the ontology*: Identification and definition of terms including concepts, verbs, instances and properties related to the process of SDM, QA, and RSDP respectively. These terms were discovered and confirmed from the industrial partners' assessment.
- *Activity 3: Concept characteristics*: We defined the following characteristics for each concept: its attributes, associations with other concepts, generalizations in which the concept is involved and constraints, if any.

We focus on elaborating the SDM ontology and showing its particularization to the QA and RSDP for supporting the objective of Q-Rapids. Some works propose using UML for ontology definition and development [11]. We use UML packages to represent ontology structure and UML class diagrams for defining classes, attributes and the relationships between them.

4   Strategic Decision-Making Ontology

4.1   Definition of the Ontology Structure

The purpose of the SDM ontology is to provide a common underlying language integrating the concepts to manage decisions and to handle strategic indicators and, consequently, it could be used as a basis for the construction of tools supporting strategic decision making. The SDM ontology is structured into two packages: Strategic Indicator and Decision. The former package groups all the concepts related to the strategic indicator concept whereas the latter package groups the concepts related to the decision concept. The < <use > > dependency between both

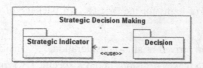

Fig. 2. SDM structure as an UML package

packages defines that the Decision package can access concepts defined in the Strategic Indicator package. Figure 2 shows the SDM structure as a UML package diagram.

4.2 Extraction of Terms Relevant for the Ontology

Tables 1 and 2 show the relevant concepts for the SDM ontology, their definition and examples. We extracted those concepts from the following sources:

- The Q-Rapids description of action (the document containing the details of how the Q-Rapids project will be carried out). We reviewed this document to find the terms representing relevant concepts in the strategic decision making. For instance, one of the objectives of the Q-Rapids project is to provide quality-related strategic indicators to support decision makers in managing the development process from a quality-aware perspective. We extracted relevant concepts as Strategic Indicator and Decision Maker from this objective.
- The partners' analysis. As mentioned above, we elaborated and generalized the terms and concept characteristics of the ontologies from the Q-Rapids industrial partners. From the answers to the semi-structured interviews provided by Q-Rapids industrial partners, we confirmed and discovered relevant concepts. For example, one respondent declared "I am also a managing director, which means that I have a role in the upper level decision part of the company". We extracted from this answer relevant concepts as Role and Decision.

Table 1. Definition of terms for the Strategic Indicator package of the SDM ontology

Concept	Definition	Examples
Strategic indicator	An aspect that a company considers relevant for the decision-making process	Customer satisfaction, product quality, blocking
Entity	Constituent part of a product and its environment for which a strategic indicator could be defined	Software product, feature
Factor	Property of an entity (or part of it) that is related to the product's quality	Maintainability, reliability
KPI	Metric that measures the degree of achievement of a strategic indicator	Percentage of user stories delivered as planned
KPI evaluation	Measurement of a KPI at a certain point of time	40% user stories delivered on time at the end of iteration 3.
Role	Position or purpose that someone has in an organization	Product manager, software developer

Table 2. Definition of terms for the `Decision` package of the SDM ontology

Concept	Definition	Examples
Decision	Determination arrived at after consideration	Hiring another developer
Decision maker	A person who makes decisions about a specific factor	Project manager
Decision rule	Rule encoding preferences of decision makers (i.e., several decision alternatives or conflicts)	When a conflict arises, quality levels will be prioritized
Constraint	Condition or restriction that affects to a decision	Developers cannot work in 2 activities the same day
External constraint	Constraint representing conditions that are out of the control of decision makers	Project budget
Internal constraint	Constraint that encodes conditions that may eventually influence decision making	Developers cannot work in 2 activities the same day
Action	Something done or performed	Include a quality requirement into the backlog

- Literature review. We reviewed the literature for identifying the concepts of strategic decision making relevant for decision makers. For example, in BSC [1], it is proposed to define KPIs to measure strategic indicators and actions that organizations should take to improve the values of their KPIs. We extracted from this reference relevant concepts as `KPI` and `Action`.

Other concepts were identified as the ontology construction process was progressing and were generalized to cover different scenarios from those of Q-Rapids.

4.3   Concept Characteristics

This section presents the UML class diagrams for both packages of the SDM ontology. Each class diagram represents the concepts, their attributes, associations with other concepts, generalizations in which the concept is involved, and constraints. All this information was extracted from the same sources detailed in the previous section.

Figure 3 shows the concepts related to the `Strategic Indicator` concept. A strategic indicator has a name (e.g. customer satisfaction), an optional description and may refine other strategic indicators forming a graph (e.g. customer satisfaction may be refined as time-to-market, product value and product quality). Strategic indicators are defined for an `Entity` by some `Roles` of the company (e.g. customer satisfaction may be defined for a specific software product by the product director) and followed by some `Roles` (e.g. customer satisfaction may be followed by sales employees). Moreover, a strategic indicator may be measured by a `KPI` (e.g. time-to market may be measured as the time it takes from defining a product to its delivery) and it is related to one or more factors (e.g. customer satisfaction may be related to usability `Factor`). `Factors` are properties of entities that may be measured. `KPI Evaluations` are assessed using metrics and evaluations associated to each strategic indicator factor at different time points for a specific `KPI`.

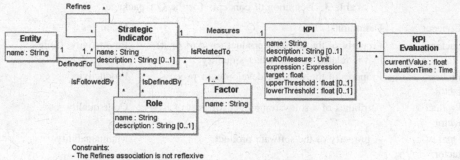

Fig. 3. UML class diagram for the strategic indicator package.

Figure 4 shows the concepts related to `Decision`. A Decision has a description and a timestamp. Decisions are made by `Decision makers` (e.g., a project manager decides to hire another developer), may consider `Decision rules` (e.g., if the percentage of delivered user stories on time is less than 40%, hiring new developers will be prioritized instead of other alternatives) and may be affected by either `External` or `Internal Constraints` (e.g., hiring new developers may be constrained by the project budget). A decision may involve one or more `Actions` (e.g. re-planning the project may involve moving out requirements from the backlog).

The SDM ontology was defined and it is being used in the context of the Q-Rapids project, but it is designed as general-purpose and therefore we argue that it may be used in other contexts, as for instance, to improve the productivity of a product chain.

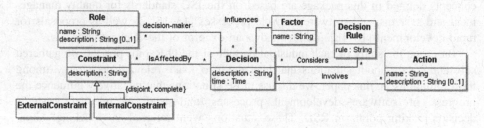

Fig. 4. UML class diagram for the decision package

5   Extending the SDM Ontology for the Case of Quality in RSD

The SDM ontology package has been extended with QA and RSDP packages (shown in Fig. 1).

Specifically, the QA package includes the concepts related to the assessment of the level of software quality during development and runtime. The concepts defined in this package are coming from the Quamoco quality meta-model [12]. Table 3 shows an

Table 3. Definition of concepts for the QA package

Concept	Definition	Examples
Factor	Property of the software product (or part of it) that is related to the product's quality	Productivity, Code quality, Reliability
Process factor	Property of the software development process	Productivity
Product factor	Attributes of a software product (or part of it)	Code quality
Quality factor	A property of the software product	Maintainability
Data source	Contains information to calculate metrics related to the software product or process	SonarQube, Jenkins, issue tracking systems
Metric	Provides a means to quantify factors that characterize an entity	Lines of code, test coverage
Entity	Constituent part of a software product	Feature, product tests
Instrument	Artifact used to determine the value of a measure, either using some tool or manually	SQALE plugin of SonarQube, FindBugs
Aggregation	Function used to aggregate values of other measures	Utility functions

excerpt of the concepts defined in this package. Factor and Entity redefine the same concepts of the SDM ontology. While they are referring to software products in QA, in the SDM ontology they are referring to any kind of products.

The RSDP package includes the concepts related to the development process focusing on the software life cycle integrating quality and functional requirements. The concepts defined in this package are based on the ISO standards for quality management and systems, and software lifecycle processes [13–15], as well as proposals for rapid development [16–18]. Table 4 shows an excerpt of the relevant concepts.

Based on the study of each industrial partner of the Q-Rapids project, we gathered substantial data about situations that lead them to make relevant decisions. Among these situations, in this paper, we discuss those situations that negatively influence the progress of software development processes and lead to relevant strategic decision-making points in RSD. These situations were referred as "blocking" situations. Reporting these situations in a strategic decision making dashboard may help decision makers to make decisions to correct those situations. In this section, we provide some insights on the use of the SDM ontology extended with the QA and RSDP packages to support the definition of blocking situations.

Blocking was defined as an SDM::Strategic Indicator. As blocking situations have detected to be relevant mainly at the level of a feature, this strategic indicator was related to the feature as an entity SDM::Entity:Feature. Based on the information gathered from different participants that played diverse roles in the software development processes, we identified several factors related to the blocking strategic indicator. For reasons of space, we will elaborate just one of these identified factors, namely testing status (i.e., QA::ProductFactor:TestingStatus).

Table 4. Definition of concepts for the RSD package

Concept	Definition	Examples
Role	Position or purpose that someone has in an organization	Product Owner
Feature	Functional or non-functional distinguishing characteristic of a system	Apply standard theme from mail theme catalogue
Process	A set of interrelated or interacting activities which transforms inputs into outputs	Requirements elicitation, feature development and testing
Quality requirement	Specify how well the system performs its intended functions	Performance, maintainability
Rapid software development	Step from agile software development that focuses on organizational capability to develop, release, and learn from software in rapid parallel cycles	Continuous delivery
Release	Describes an increment into complete software product valuable to customers	Versions (e.g. Version 1.0 registration and login management)
Sprint	Short time frame, in which a set of software features is developed, leading to a working product	One week sprint
Project backlog	The user stories the team has identified for implementation	Collection of user stories
User story	Simple narrative illustrating the user goals that a software function will satisfy	As a < tester > , I want to < apply the first prototype standard desktop theme > so that I can < provide feedback on the concept>

We elicited the definition of the testing status factor based on the information provided by some roles. The testing status factor was mentioned among others by test managers as a situation that affects the blocking strategic indicator. For instance, negative integration testing results might block the release of a feature. Therefore, using the RSDP package we instantiated RSD:Role:TestManager.

Testing status was quantified using different metrics, for example: QA::Metric:TestCoverage and QA::Metric:NumberOfTestsDone. To gather information for such metrics, different modules from continuous integration tools (e.g., Jenkins) or continuous code quality tools (e.g. SonarQube) were used. Therefore, we defined QA::DataSource:SonarQube as the data source containing the information to calculate the first metric and QA::DataSource:Jenkins for the second one.

The degree of achievement of the blocking strategic indicator would be defined as an aggregation of values coming from the diverse factors that defines it. Hence, once these aggregations (e.g., AG1, ..., AGn) have been defined, the corresponding SDM:

`KPI:BlockingKPI` can be defined as well. Furthermore, decisions related to blocking can be taken by diverse roles as, for instance, a product manager that given a blocking situation raised by testing status issues on a specific feature, decides DC1: "to postpone the feature to the next release after agreeing that with the corresponding client". Such decision is based on an existing decision rule, DR1 : "if the entity causing the testing status issue does not have a high impact on the final product, then it could be postponed to future releases" respecting the internal constraint IC1: "Entity releases must be 100% agreed with the client". So, `SDM::Role:ProductManager`, `SDM::Constraint:InternalConstraint:IC1`, `SDM::Decision:DC1`, and `SDM::DecisionRule:DR1` are instances of the SDM ontology.

6 Conclusions

Organizations make decisions every single day to follow their strategies and to achieve their goals. This paper has presented the SDM ontology to support different strategic decision making processes and extended the ontology to cover the context of managing quality in RSD projects.

As a future work, we are planning to refine the SDM ontology extending it with new concepts to track the reasons of the decisions, to manage the prediction of violations of the strategic indicators and to explore corrective actions in the solution space through what-if-analysis. Moreover, we are going to evaluate the SDM ontology in the context of Q-Rapids industrial partners.

Acknowledgments. This work is a result of the Q-Rapids project, which has received funding from the European Union's Horizon 2020 research and innovation program under grant agreement N° 732253. We thank to Q-Rapids industrial partners for participating in our empirical studies to develop the presented ontologies.

References

1. Kaplan, R.S., Norton, D.P., Dorf, R.C., Raitanen, M.: The Balanced Scorecard: Translating Strategy into Action, vol. 4. Harvard Business School Press, Boston (1996)
2. Object Management Group: Business Motivation Model (BMM) 1.3 (2015). http://www.omg.org/spec/BMM/1.3. Accessed Apr 2017
3. Barone, D., Mylopoulos, J., Jiang, L., Amyot, D.: The Business Intelligence Model: Strategic Modelling (Version 1.0). ftp://ftp.cs.toronto.edu/csrg-technical-reports/607/BIM-TechReport.pdf. Accessed July 2017
4. Diamantini, C., Potena, D., Storti, E., Zhang, H.: An ontology-based data exploration tool for key performance indicators. In: Meersman, R., Panetto, H., Dillon, T., Missikoff, M., Liu, L., Pastor, O., Cuzzocrea, A., Sellis, T. (eds.) OTM 2014. LNCS, vol. 8841, pp. 727–744. Springer, Heidelberg (2014). doi:10.1007/978-3-662-45563-0_45
5. Maté, A., Zoumpatianos, K., Palpanas, T., Trujillo, J.C., Mylopoulos, J., Koci, E.: A systematic approach for dynamic targeted monitoring of KPIs. In: Proceedings of the 24th Annual International Conference on Computer Science and Software Engineering, pp. 192–206. IBM Corp. (2014)

6. Maté, A., Trujillo, J.C., Mylopoulos, J.: Specification and derivation of key performance indicators for business analytics: A semantic approach. Data Knowl. Eng. **108**, 30–49 (2017)
7. Object Management Group: Decision Model and Notation (DMN) 1.1 (2016). http://www.omg.org/spec/DMN/1.1/PDF/. Accessed Apr 2017
8. Guzmán, L., Oriol, M., Rodríguez, P., Franch, X., Jedlitschka, A., Oivo, M.: How can quality awareness support rapid software development? – a research preview. In: Grünbacher, P., Perini, A. (eds.) REFSQ 2017. LNCS, vol. 10153, pp. 167–173. Springer, Cham (2017). doi:10.1007/978-3-319-54045-0_12
9. Basili, V., et al.: Aligning Organizations Through Measurement - The GQM + Strategies Approach. Springer, Heidelberg (2014)
10. Fernández-López, M., Gómez-Pérez, A., Juristo, N.: METHONTOLOGY: From ontological art towards ontological engineering. In: AAAI-97 Spring Symposium Series. Stanford University, USA, 24–26 March 1997
11. Cranefield, S.: Networked knowledge representation and exchange using UML and RDF. J. Digit. Inf. 1(8) (2001). http://jodi.ecs.soton.ac.uk
12. Wagner, S., Lochmann, K., Winter, S., Deissenboeck, F., Juergens, E., Herrmannsdoerfer, M., Heinemann, L., Kläs, M., Tendowicz, A., Heidrich, J., Ploesch, R., Goeb, A., Koerner, C., Schoder, K., Streit, J., Schubert, C.: The Quamoco quality meta-model. Technical Report TUM-I1281, Technische Universität München
13. International Standardization Organization/International Electrotechnical Commission. 9000: 2005. Quality management systems-Fundamentals and vocabulary (2005)
14. International Standardization Organization/International Electrotechnical Commission. 12207: 2008. Systems and software engineering–Software life cycle processes (2008)
15. International Standardization Organization/International Electrotechnical Commission. 26515. Systems and software engineering - Developing user documentation in an agile environment. First edition 01 December 2011, Corrected version 15 March 2012 (2012)
16. Mäntylä, M.V., Adams, B., Khomh, F., Engström, E., Petersen, K.: On rapid releases and software testing: a case study and a semi-systematic literature review. Empirical Softw. Eng. **20**, 1384 (2015). doi:10.1007/s10664-014-9338-4
17. Fitzgerald, B., Stol, K.J.: Continuous software engineering: A roadmap and agenda. J. Syst. Softw. **123**, 176–189 (2017)
18. Leffingwell, D.: Agile Software Requirements: Lean Requirements Practices for Teams, Programs, and the Enterprise. Addison-Wesley Professional (2011)

Detecting Bad Smells of Refinement in Goal-Oriented Requirements Analysis

Keisuke Asano, Shinpei Hayashi[✉], and Motoshi Saeki

Department of Computer Science, Tokyo Institute of Technology,
Ookayama 2–12–1–W8–83, Meguro-ku, Tokyo 152–8552, Japan
{k_asano,hayashi,saeki}@se.cs.titech.ac.jp

Abstract. Goal refinement is a crucial step in goal-oriented requirements analysis to create a goal model of high quality. Poor goal refinement leads to missing requirements and eliciting incorrect requirements
as well as less comprehensiveness of produced goal models. This paper
proposes a technique to automate detecting *bad smells* of goal refinement,
symptoms of poor goal refinement. Based on the classification of poor
refinement, we defined four types of bad smells of goal refinement and
developed two types of measures to detect them: measures on the graph
structure of a goal model and semantic similarity of goal descriptions.
We have implemented a support tool to detect bad smells and assessed
its usefulness by an experiment.

Keywords: Goal-oriented requirements analysis · Goal refinement ·
Smell detection

1 Introduction

Goal-oriented requirements analysis (GORA) is one of the popular techniques
to elicit requirements to business processes, information systems, and software
(simply, systems hereafter), and is being made into practice and worked into university curriculums [9,23]. In GORA, customers' needs are modeled as goals to
be achieved finally by software-intensive systems that will be developed, and the
goals are decomposed and refined into a set of more concrete sub goals. After fin-
ishing goal-oriented analysis, the analyst obtains an acyclic (cycle-free) directed
graph called *goal graph*. Its nodes express goals to be achieved by the system
that will be developed, and its edges represent logical dependency relationships
between the connected goals. More concretely, a goal can be refined into sub
goals, and the achievement of the sub goals contributes to its achievement. We
have two types of goal decomposition; one is AND decomposition, and the other
is OR. In AND decomposition, if all of the sub goals are achieved, their parent
goal can be achieved or satisfied. On the other hand, in OR decomposition, the
achievement of at least one sub goal leads to the achievement of its parent goal.

According to the textbook by van Lamsweerde [9], "a *goal* is a prescriptive
statement of intent that the system should satisfy through the cooperation of its

© Springer International Publishing AG 2017
S. de Cesare and U. Frank (Eds.): ER 2017 Workshops, LNCS 10651, pp. 122–132, 2017.
https://doi.org/10.1007/978-3-319-70625-2_12

system components (agents)", while "a requirement is a goal under the responsibility of a single system component ... ". It means that requirements are derived by goal refinement, and poor goal refinement leads to missing requirements and to eliciting incorrect ones. In addition, it may cause less comprehensiveness of a goal model, and as a result communication gaps among stakeholders occur. Goal refinement is really most crucial step in GORA, and it is difficult even for experienced analysts. It is significant to detect poor goal refinement during the development of a goal model.

Figure 1 illustrates a part of a goal model of a book order system, which includes poor goal refinement. The root goal "Fulfill a book order" is refined into two sub goals with AND decomposition, which is indicated with an arc lying across the edges to the sub goals. The sub goal "Handle a receipt" does not seem to directly contribute to the achievement of its parent goal "Deliver books", rather some sub goals related to a carriage service of books, such as express delivery and designating the date of its arrival, seem to be missing. Although "Handle a receipt" may be one of the necessary functions for the system to be developed, this poor goal refinement leads to missing requirements to the achievement of "Deliver books". Moreover, generally speaking, sub goals which do not contribute to the achievement of a root goal lead to incorrect requirements, irrelevant to customers' needs. In this case, we can guess the poor refinement from the symptoms or signs that there is only one sub goal and that "Deliver books" is not so semantically related to "Handle a receipt". We have to detect *bad smells* of poor goal refinement, symptoms or signs that indicate a potential of a problem against requirements elicitation of high quality, during developing a goal model.

This paper proposes an automated technique to detect bad smells of poor goal refinement. As discussed above, we use the measures related to (1) structural characteristics of a goal model as a graph to detect goal refinement having only one sub goal and (2) similarity between goal descriptions of a parent goal and its sub goals to detect their semantic relation. We develop several metrics to detect bad smells of goal refinement. Note that this work was done in Japanese, and all the example goal graphs written in English used in the studies were translated into Japanese to apply our tool to them.

The rest of this paper is organized as follows. Sections 2 and 3 present the developed smell detection techniques and the experimental evaluation of the techniques, respectively. Sections 4 and 5 are for related work and concluding remarks.

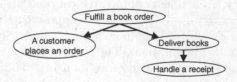

Fig. 1. Example of a goal graph having poor refinement.

2   Detecting Bad Smells in Goal Refinement

Based on our experiences, we extracted three categories of poor goal refinement: (1) a sub goal does not contribute to the achievement of its parent directly or is irrelevant to the achievement of its parent, (2) there is a lacking sub goal to achieve its parent, and (3) a leaf goal is not concrete. The following subsections explain how these symptoms can be detected using structural and semantic metrics. We introduce four types of bad smells of goal refinement. Low Semantic Relation and Many Siblings are defined for detecting poor refinement of the first category. Few Siblings is desinged for the second category, and Coarse Grained Leaf is for the last one.

2.1   Detecting Low Semantic Relation

The semantic similarity between goals are calculated with their goal descriptions written in natural language, and therefore we use a lightweight natural language processing technique, the case frame approach [3].

Case Frame.   The technique we adopted is based on case frames of the Fillmore's case grammar, which are semantic representations of natural language sentences [3]. The technique based on the case frame approach has been used widely in requirements engineering [13,14,16]. A case frame consists of a verb and semantic roles of the words that frequently co-occur with the verb. These semantic roles are specific to a verb and are called *case* (precisely, *deep case*). For example, "Deliver a book" is transformed into the case frame (Deliver, $^{actor:}$ _, $^{object:}$ Book, $^{destination:}$ _). Actor, object, and destination are case names and "–" stands for no words filled in the slot. Generally speaking, a verb has multiple meaning, so to identify its meaning in a goal description, we use constraints on the words co-occurring with it. For example, "deliver" has the meaning of "speak", e.g., delivering a speech, in addition to "convey". When "deliver" is used as the meaning of "speak", the case slot of "object" should be filled with words denoting something abstract concepts such as "speech", "lecture", not words denoting the concepts of physical goods like "book". We use this kind of constraints on the semantic concepts of words that case slots can be filled with. Therefore we have a dictionary of case frames and that of hierarchical (semantic) concepts. This approach is the same as the work by Nakamura *et al.* [11].

Semantic Similarity.   A dictionary of hierarchical concepts is used to calculate semantic similarity between goal descriptions. Suppose that we have three goal descriptions: "Deliver a book", "Handle a receipt", and "Send a receipt" and obtain their case frame representations. The occurrences of words in these case frames are mapped into their semantic concepts using the dictionary of hierarchical concepts, and the similarity measures among these words are calculated. The similarity measure between the word A and B is the relative distance from their common ancestor to them in the hierarchy of the concepts [20].

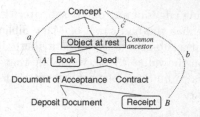

Fig. 2. Similarity of words in hierarchical structure.

Figure 2 illustrates how to calculate a similarity measure between the concepts "Book" and "Receipt". In the figure, the graph expresses a hierarchical structure of concepts. For example, the concepts "Concept" and "Receipt" are a root concept and a child of "Document of Acceptance", respectively. Then, the similarity measure between A and B is calculated as $Similarity_{word}(A, B) = (c + c)/(a + b)$ where a and b are the depth of the concepts A and B from the root concept respectively, while c is the depth of the common ancestor of A and B. If A and B are the same, i.e., the two words denote the same concept, $Similarity_{word}(A, B) = 1$, the maximum value. If the common ancestor of A and B is the root, i.e., there are no semantic intersection between them, the value is the minimum, 0.

Next, we define how to measure the semantic similarity of goal descriptions using this metric. After transforming a goal description in a case frame, we obtain the concepts of the words with which the slots of the case frame are filled, using the dictionary of hierarchical concepts. We calculate $Similarity_{word}$ to the words appearing in the goal descriptions. In the above example, using the dictionary, we can obtain $Similarity_{word}(\mathsf{Deliver}, \mathsf{Handle}) = 0.53$ and $Similarity_{word}(\mathsf{Deliver}, \mathsf{Send}) = 0.93$. Thus from the viewpoint of verbs, i.e., activities, the description "Send a receipt" is semantically closer to "Deliver a book" rather than "Handle a receipt". Let $cases(g)$ be a set of the concepts in the case frame representation of the description of the goal g, e.g., $cases(\text{"Deliver a book"}) = \{\mathsf{Deliver}, \mathsf{Book}\}$. Thus, the following predicate is defined for deciding if there are semantic relationships between the goals connected with the edge e or not: $Similarity(g_1, g_2) = \max_{w_1 \in cases(g_1), w_2 \in cases(g_2)} Similarity_{word}(w_1, w_2)$. Intuitively speaking, we calculate $Similarity_{word}$ between the words appearing in the descriptions of the parent goal and its sub goal connected with the edge e. The smell is detected if these goals have less semantic similarity; its value is less than a certain threshold $Th_{similarity}$.

2.2 Detecting Many Siblings and Few Siblings

It is easy to calculate the number of sibling edges of an edge. $SiblingEdges(e)$ counts the sibling edges of e, including e itself, and it is defined as $SiblingEdges(e) = |\{e' \mid s(e) = s(e')\}|$ where $s(e)$ is a parent goal connected with e.

If the number of sibling edges of an edge e is greater than a certain threshold Th_{high}, we decide that e has the bad smell of Many Siblings, which contributes to find poor refiment that a sub goal does not contribute directly or is irrelevant. If the number is less than the value Th_{low}, we decide that e has the bad smell of Few Siblings, which leads to finding poor refinement that there is a lacking sub goal as a sibling edge of e.

2.3   Detecting Coarse Grained Leaf

As for the granularity of leaf goals, we calculate the relative depth of a leaf from a root goal in a graph. See Fig. 3. This example has five leaf goals A, C, D, F, and G. According to van Lamsweerde [9], each of these leaf goals is achieved by a single agent and leads to a requirement that the system to be developed will realize as a function. If they are not sufficiently refined yet, they are vague, and it is difficult to realize them as concrete functions of the system. The depth or distance of the leaf goal A from a root is smaller rather than those of the other leaves, in particular than F and G, and it can be considered that A is necessary to be refined further, comparing with F and G. Thus, we detect the smell Coarse Grained Leaf by focusing on the metric measuring relative depth of goals, and the measure we use has been based on the idea of combining reachability probability and Shannon's information entropy, proposed by Shao [17]. The reason why we did not use simple measures such as the distance of the goal from a root goal is that the factor of branch breadth as a graph should be considered.

We followed this idea, *equitability* of probabilities of reaching leaf nodes, and adapted it to the measure that can express that a leaf has a depth from a root node on the same level with other leaves. They used the idea of probability to reach a leaf from a root and Shannon's information entropy. Based on it, we consider that all of the leaf nodes are on the same level of depth or have the same granularity when their reachability probabilities from a root to leaves are the same. Under the assumption that each sibling edge has an equal probability at branching, i.e., the probabilities of reachability from a parent goal to its sub goals are the same, we can have a reachability probability of the edge e as 1 div $SiblingEdges(e)$. In the example of Fig. 3, the reachability probabilities from A to B and from B to D are $1/2$ and $1/3$, respectively. Let $\langle e_i^g \rangle_{i=0,...,n}$ be the edge sequence from a root goal to a goal g. We can define the reachability probability from a root to g as $Reachability(g) = \prod_{e \in \langle e_i^g \rangle_{i=0,...,n}} 1/SiblingEdges(e)$. In the example of Fig. 3, we obtain $Reachability(A) = 1/2$ and $Reachability(F) = 1/2 \cdot$

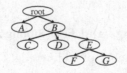

Fig. 3. Granularity of leaf goals.

$1/3 \cdot 1/2 = 1/12$. We defined an expression to identify leaf goals having coarse-grained granularity, i.e., having the possibility to be refined further comparing with the others. If leaf goals are the result of uniform refinement, i.e., they have the same granularity, their reachability probability is also the same, i.e., 1 div m where m is the number of the leaf goals. In order to detect the smell, we measure the ratio of the reachability probability of a goal to the *ideal* one, i.e., $1/m$: $Reachability(g)/(1/m) = m \cdot Reachability(g)$. If this value is greater than a certain threshold Th_{reach}, we detect Coarse Grained Leaf on the goal g as it is not refined well rather than the others. In the example of Fig. 3, the goals A and F have $(1/2) \cdot 5 = 2.5$ and $(1/12) \cdot 5 \simeq 0.42$, respectively. In this way, we can decide that the goal A is coarser-grained than the others and should be refined further.

3   Experimental Evaluation

We have implemented an automated tool supporting our technique as a plug-in of our already developed tool for the attributed goal-oriented analysis [15]. For analyzing goal descriptions and calculating their similarity, it uses a part of the EDR electronic dictionaries[1], which includes general-purpose Japanese dictionaries of case frames and hierarchical concepts.

The aim of the experiment is to evaluate if our approach and the tool can indicate bad smells, which are symptoms of problems for requirements elicitation. Our experiment is comparative where we compare the correct set of poor refinement with the results of the indicated bad smells by the tool. However, it is difficult to create complete and real correct set of poor refinement because the judgment of poor or not poor refinement is subjective to human analysts. Although it is difficult to have a complete correct set, we can have an approximate correct set that several analysts agree with, and we considered it as a correct set. Thus, our experimental procedure includes how to construct a correct set of poor refinement. This correct set may include errors, i.e., the occurrences of non-poor refinement and missing poor refinement. So we should check if our technique can detect these errors if any.

3.1   Experimental Procedure

Figure 4 shows the overview of our experiment. We had four subjects which had experienced in GORA to detect poor goal refinement in examples of goal models to check the results of indicating bad smells by our tool. They were given two goal models: Mobile Personal Emergency Response System (MPER) [10] and Museum-Guide System (MGS) [1], and were independently required to indicate the parts of poor goal refinement that may cause the problems of requirements elicitation. They were also required to express explicitly the rationales why the indicated parts were poor refinement. After performing this independent task,

---

[1] https://www2.nict.go.jp/out-promotion/techtransfer/EDR/.

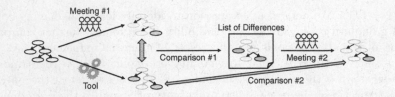

Fig. 4. Flow of the experimental evaluation.

they had a face-to-face meeting (Meeting #1) to get a consensus of their indicated results that were different from each other. Their concerted results could be considered as the first version of a correct set of poor refinement. We compared it with the bad smells that our tool indicated (Comparison #1), and its result is shown as "List of Differences" in the figure. The threshold values that we used in this experiment were $(Th_{similarity}, Th_{low}, Th_{high}, Th_{reach}) = (0.7, 2, 5, 2)$, which were empirically decided based on the analysis of existing goal models.

We had a postprocess to refine the obtained first version of the correct set. Our tool indicated edges and leaf goals only while our subjects could do any parts in the goal models. For example, the subject indicated a goal as poor refinement, but the tool did the edge incoming to the goal indicated by them. Although the parts that the subjects and tool indicated are different, we should check if they indicated the same refinement or not. In case that such a discordance was found and/or that the subjects missed correct indications or identified wrong indications, the second face-to-face meeting (Meeting #2) was conducted by the same subjects to confirm their indications by comparing them with the tool's ones and to fix them if necessary. One of the authors facilitated this meeting, but he just suggested to the subjects the parts that they should discuss. After this meeting, we have got the correct set modified by the subjects. We compared it with the bad smells that the tool indicated (Comparison #2). We measured precision and recall values by comparing the poor refinements that the tool detected and those indecated by subjects.

3.2   Results and Discussion

Table 1 shows the number of the goals for each goal model (# Goals), the number of indications by the tool (# Tool indicated), the number of the correct set, i.e., indications by the subjects (# Correct set), and the number of the indications same as the subjects, i.e., the correct indications by the tool (# The same as the correct set), and precision and recall values. In the example of MPER, our tool indicated 28 bad smells, and our subjects agreed that there were 11 occurrences of poor refinement.

As for the errors of the correct set, it included seven and four occurrences of missing poor refinement in MPER and MGS respectively. Our subjects did not find them during Meeting #1 of Fig. 4, however after checking the output of the tool, they noticed them as poor refinement during Meeting #2. It means that our tool can find poor refinement which human analysts might miss.

Table 1. Detection results

	MPER	MGS	Total
# Goals	35	37	72
# Tool indicated ($|T|$)	28	21	49
# Correct set (subjects indicated) ($|C|$)	20	17	37
# The same as the correct set ($|T \cap C|$)	11	11	22
Precision ($|T \cap C|/|T|$)	0.39	0.52	0.45
Recall ($|T \cap C|/|C|$)	0.55	0.65	0.59

As shown in Table 1, the precision value of MPRE is 0.39, less than 50%. Its major reason is that our technique did not consider domain-specific knowledge. To calculate the semantic similarity between a parent goal and its sub goals, we used the knowledge on general concepts, which resulted from the general dictionary of words in EDR. In the example of MPER, a parent goal "Emergency is detected" was refined into "Vital signs are processed". The word "vital sign" is domain-specific and is not included in our dictionary. Thus, our tool failed in deciding the existence of the semantic relationship between these goals. On the other hand, since MGS was more popular than MPER, it led to higher precision.

The recall values were more than 50% in both of the goal models. The major reason is similar to the case of the precision values, i.e., semantic similarity of goal descriptions. Although our subjects indicated the occurrences of the sub goals semantically less related to their parents, the tool decided that they were similar. Our metric on semantic similarity should be more elaborated. Another reason is that our technique did not consider contexts associated with goal refinement. In the example of MPER, our subjects indicated that the refinement of "Analyse vital signs" into "Detects emergency" was poor because there was the goal "Emergency is detected" as an ancestor goal, i.e., almost the same goal description as that of the sub goal, while the tool did not. Although the meaning of the sub goal was "Detects emergency *with sensed data of vital sign*" in the context of its parent goal "Analyse vital signs", our subjects indicated that this leaf goal was not so concrete and suggested to improve its goal descriptions or to refine it further. We need to develop semantic processing of goal descriptions considering such a context, e.g., considering not only a parent goal and its *direct* sub goals but also its ancestors and descendants.

3.3   Threats to Validity

External Validity. External validity is related to the generality of the obtained conclusions. In this experiment, although we used only two goal models, their problem domains were different, and we believe that they covered varieties of the problem domain in a certain degree. The sizes of the goal models were smaller than those in practical levels. Our technique detected real 22 occurrences, and it means that there may be many occurrences in spite of smaller sizes of goal models. Our tool can work well on the smaller sizes of models. The scalability of our approach, e.g., performance, should be checked as one of future research agenda.

Construct Validity. This validity shows that our observed data can really reflect our obtained conclusions. We assessed our technique by the differences between the correct set and the occurrences the tool indicated. The possibility of this threat is the quality of the correct set. The correct set was made by our subjects, so its quality depended on their skills and knowledge. Another factor is the quality of the translations of English goal descriptions into Japanese, and it may lead them to misreading. However, we took four subjects experienced in goal modeling, and any decisions resulted from their consensus. So, we consider the quality of the correct set was sufficiently high.

Internal Validity. We should explore the factors of affecting the obtained results other than those of our approach. One of the possibilities of these factors is the bias at Meeting #2 in Fig. 4 that our subjects might have taken their decisions advantageous to our tool. However, they rejected many occurrences ($27 = 49 - 22$ as shown in Table 1) that the tool indicated as poor refinement, and this fact shows the fairness of our subjects. In addition, we took a process of agreement of all subjects to change their decisions.

4   Related Work

There are wide varieties of studies to assist in improving the quality of requirements models, including goal models. Kaiya et al. [7] proposed the measures of quality characteristics and they attach some attribute values such as contribution and preference values attached to goal models. By using the values, the proposed technique measures the correctness completeness, unambiguity, etc. of goal models. This technique can be extended to reason the changeability of goal models [6, 19]. Giorgini et al. [4] developed two measures contribution and denial values of goal achievement, to measure the quantitative quality of goal models. On goal models written in i^* and KAOS, the metrics of their complexity were also developed [2,5], and almost all of them calculates the numbers of syntactic components of goal models such as agents, edges, etc. Silva et al. [18] proposed visualization quality of i^* goal models to improve their understandability. As mentioned above, although there are several work to define metrics of quality and complexity of goal models from various viewpoints, there are none of the metrics on quality of goal refinement from structural and semantical viewpoints, and they were not for detecting poor refinement. As for quality from a semantical viewpoint, note that Kaiya et al. [8] proposed metrics of correctness, completeness, unambiguity, etc. for requirements written in a natural language using mapping words into domain ontology as semantic concepts. To enhance semantic similarity in our approach, using domain ontology is one of the promising approaches.

The second category is the support of goal refinement of higher quality. Refinement patterns can guide goal refinement in KAOS [9]. They are on logical basis of temporal logic, and they are not for detecting poor refinement. More concretely, refinement following the suitable refinement patterns can guarantee logically achievement of a parent goal by sub goals, but they did not consider the other aspects of quality such as the granularity of refined goals. Our approach focused on bad smell detection of poor goal refinement and not on the improvement. The idea of refinement patterns is complementary to ours when we make a catalog of bad smells and their resolutions as patterns.

The last category is related to the quality of natural language sentences. Our approach did not evaluate goal descriptions themselves, but their semantic relationships. To measure the quality of goal descriptions, we should adopt natural language techniques to analyze ambiguity and vagueness of natural language sentences. Yang *et al.* [21,22] discussed nocuous ambiguity such as the scope of conjunctives "and" and "or", and automated to detect its occurrences. They applied their approach to usual natural language sentences, not goal descriptions. Goal descriptions are short sentences and may be incomplete. Although their approach is complementary to ours, we cannot apply it as it is. We need some techniques to supplement descriptions with adding missing words [12] and then apply it to the supplemented descriptions.

5   Conclusion

This paper discussed the automated technique to detect the occurrences of poor refinement in a goal model. After clarifying bad smells of poor refinement, we defined two types of measures: structural characteristics of a goal model as a graph and semantic aspects of goal descriptions. Our experimental results showed that our direction is promising and some improvement is necessary.

Future work can be listed up as follows: (1) more experimental analyses on goal models of wide varieties of problem domains and in practical level, (2) enhancing metrics, in particular for detecting bad smells of the first category, i.e., a sub goal does not contribute directly or is irrelevant, including more elaborated natural language processing, and (3) technique to support the improvement of detected bad smells, so-called *refactoring* of goal refinement.

Acknowledgments. This work was partly supported by JSPS Grants-in-Aid for Scientific Research Nos. JP15K00088, JP15K15970, and JP15H02685. We would like to thank Prof. Takako Nakatani for giving us sample goal models having poor refinement.

References

1. Ali, R., Dalpiaz, F., Giorgini, P.: A goal-based framework for contextual requirements modeling and analysis. Requir. Eng. **15**(4), 439–458 (2010)
2. Espada, P., Goulão, M., Araújo, J.: A framework to evaluate complexity and completeness of KAOS goal models. In: Salinesi, C., Norrie, M.C., Pastor, Ó. (eds.) CAiSE 2013. LNCS, vol. 7908, pp. 562–577. Springer, Heidelberg (2013). doi:10. 1007/978-3-642-38709-8_36

3. Fillmore, C.J.: Lexical entries for verbs. Found. Lang. **4**, 373–393 (1968)
4. Giorgini, P., Mylopoulos, J., Nicchiarelli, E., Sebastiani, R.: Reasoning with goal models. In: Spaccapietra, S., March, S.T., Kambayashi, Y. (eds.) ER 2002. LNCS, vol. 2503, pp. 167–181. Springer, Heidelberg (2002). doi:10.1007/3-540-45816-6_22
5. Gralha, C., Araújo, J., Goulão, M.: Metrics for measuring complexity and completeness for social goal models. Inf. Syst. **53**, 346–362 (2015)
6. Hayashi, S., Tanabe, D., Kaiya, H., Saeki, M.: Impact analysis on an attributed goal graph. IEICE Trans. Inf. Syst. **E95–D**(4), 1012–1020 (2012)
7. Kaiya, H., Horai, H., Saeki, M.: AGORA: attributed goal-oriented requirements analysis method. In: Proceedings of the RE, pp. 13–22 (2002)
8. Kaiya, H., Saeki, M.: Ontology based requirements analysis: lightweight semantic processing approach. In: Proceedings of the QSIC, pp. 223–230 (2005)
9. van Lamsweerde, A.: Requirements Engineering: From System Goals to UML Models to Software Specifications. Wiley, Hoboken (2009)
10. Mendonça, D., Ali, R., Rodrigues, G.N.: Modelling and analysing contextual failures for dependability requirements. In: Proceedings of the SEAMS, pp. 55–64 (2014)
11. Nakamura, R., Negishi, Y., Hayashi, S., Saeki, M.: Terminology matching of requirements specification documents and regulations for compliance checking. In: Proceedings of the RELAW, pp. 10–18 (2015)
12. Negishi, Y., Hayashi, S., Saeki, M.: Supporting goal modeling for eliciting regulatory compliant requirements. In: Proceedings of the CBI, pp. 434–443 (2017)
13. Ohnishi, A.: Software requirements specification database based on requirements frame model. In: Proceedings of the ICRE, pp. 221–228 (1996)
14. Rolland, C., Proix, C.: A natural language approach for Requirements Engineering. In: Loucopoulos, P. (ed.) CAiSE 1992. LNCS, vol. 593, pp. 257–277. Springer, Heidelberg (1992). doi:10.1007/BFb0035136
15. Saeki, M., Hayashi, S., Kaiya, H.: A tool for attributed goal-oriented requirements analysis. In: Proceedings of the ASE, pp. 670–672 (2009)
16. Saeki, M., Kaiya, H.: Supporting the elicitation of requirements compliant with regulations. In: Bellahsène, Z., Léonard, M. (eds.) CAiSE 2008. LNCS, vol. 5074, pp. 228–242. Springer, Heidelberg (2008). doi:10.1007/978-3-540-69534-9_18
17. Shao, K.T.: Tree balance. Syst. Biol. **39**(3), 266–276 (1990)
18. Silva, L., Moreira, A., Araújo, J., Gralha, C., Goulão, M., Amaral, V.: Exploring views for goal-oriented requirements comprehension. In: Comyn-Wattiau, I., Tanaka, K., Song, I.-Y., Yamamoto, S., Saeki, M. (eds.) ER 2016. LNCS, vol. 9974, pp. 149–163. Springer, Cham (2016). doi:10.1007/978-3-319-46397-1_12
19. Tanabe, D., Uno, K., Akemine, K., Yoshikawa, T., Kaiya, H., Saeki, M.: Supporting requirements change management in goal oriented analysis. In: Proceedings of RE, pp. 3–12 (2008)
20. Wu, Z., Palmer, M.: Verbs semantics and lexical selection. In: Proceedings of ACL, pp. 133–138 (1994)
21. Yang, H., De Roeck, A.N., Gervasi, V., Willis, A., Nuseibeh, B.: Extending nocuous ambiguity analysis for anaphora in natural language requirements. In: Proceedings of the RE, pp. 25–34 (2010)
22. Yang, H., Willis, A., De Roeck, A.N., Nuseibeh, B.: Automatic detection of nocuous coordination ambiguities in natural language requirements. In: Proceedings of the ASE, pp. 53–62 (2010)
23. Yu, E.S.: Social modeling and i^*. In: Borgida, A.T., Chaudhri, V.K., Giorgini, P., Yu, E.S. (eds.) Conceptual Modeling: Foundations and Applications. LNCS, vol. 5600, pp. 99–121. Springer, Heidelberg (2009). doi:10.1007/978-3-642-02463-4_7

Requirements Engineering for Data Warehouses (RE4DW): From Strategic Goals to Multidimensional Model

Azadeh Nasiri[1,2], Waqas Ahmed[1,2(✉)], Robert Wrembel[1], and Esteban Zimányi[2]

[1] Institute of Computing Science, Poznan University of Technology, Poznań, Poland
robert.wrembel@cs.put.poznan.pl
[2] Department of Computer and Decision Engineering, Université Libre de Bruxelles, Brussels, Belgium
{nazadeh,waqas.ahmed,ezimanyi}@ulb.ac.be

Abstract. Business Intelligence (BI) systems help organisations to monitor the fulfillment of business goals by means of tracking various Key Performance Indicators (KPIs). Data Warehouses (DWs) supply data to compute KPIs and therefore, are an important component of any BI system. While designing a DW to monitor KPIs, the following two important questions arise: (1) What data should be stored in a DW to measure a KPI, and (2) how the data should be modelled in a DW? We present a model-based Requirement Engineering (RE) framework to answer these questions. Our proposal consists of two major modelling components, namely, the context modelling component which is used to represent why and which data is required, and data modelling component which is used to model data as a multidimensional model.

Keywords: Data warehouse · Requirements engineering · Key performance indicators · Multidimensional model

1 Introduction

Business Intelligence (BI) systems enable decision-makers to analyse the business data, facilitate the decision-making, and improve business performance. One of the main applications of BI systems is to track Key Performance Indicators (KPIs). KPIs are complex measurements used to present the achievement of business strategies in a user-friendly format. Data warehouses (DWs) are one of the main components of any BI system and supply data to compute KPIs. They integrate data from disparate, heterogeneous, organization-wide systems and often represent them in a multidimensional (MD) model. Often, oganizations find it challenging to utilize KPIs to monitor business goals because underlying DWs are designed with no connection to business strategies and are incapable of supplying the required data [6]. This causes many BI projects to fail.

© Springer International Publishing AG 2017
S. de Cesare and U. Frank (Eds.): ER 2017 Workshops, LNCS 10651, pp. 133–143, 2017.
https://doi.org/10.1007/978-3-319-70625-2_13

In general, RE is the process of discovering the needs of stakeholders and supporting those needs by the means of communicable models and documents. RE approaches for designing DWs deal with the questions of what data is of particular interest for decision-makers, and in which form it should be stored in a DW [5,12]. These approaches can be classified into three groups [15], namely, supply driven, demand driven, and hybrid approaches. The *supply-driven* approaches obtain an MD model from the detailed analysis of data sources. Being too much data-oriented, such approaches are unable to capture the business relevance of the data. In *demand-driven* and *hybrid* approaches, a DW design is driven by business requirements. The existing RE methods based on demand-driven approaches use various drivers such as business processes [2,3,7], decision process [16], and goals and objectives of an organization [1,4,5,9,11] to determine data elements to be stored in DWs. There are few proposals [6] that show how KPIs can be linked to organisational goals, and computed as queries from a DW but lack to address what organisational data supports monitoring KPIs.

In this paper, we propose a dedicated RE framework named RE4DW to provide the settings for the structured and efficient communication among stakeholders. To this end, we have adopted a model-based approach in which models provide the basis of requirements artifacts. RE4DW is composed of two modelling components: (1) *context modeling* using the i* framework, which answers what organisational data to be stored in a DW that supports monitoring KPIs, and (2) *data modeling* using an MD model, which answers how to structure this data. RE4DW includes a set of steps that systematically derives an MD model from KPIs by which the achievement of organisational goals are measured. KPIs are initially captured in the context modeling and then through some steps proposed by RE4DW are broken into the data to be stored in a DW. Finally, the data are structured in a MD model in the data modeling. This paper is the extension of our previous work presented at [12,13] which involves using the data sources to design a DW for monitoring KPIs. Besides, RE4DW benefits of being applied to a real case study from the energy sector. The rest of this paper is organized as follows: In Sect. 2, we elaborate the two modelling components of the RE4DW framework and provide the steps to build a goal model and an MD model starting from high-level strategic goals. In Sect. 3, we illustrate RE4DW with a case study from the energy sector. Section 4 we review the related work and in Sect. 5, we conclude this paper.

2   RE4DW Framework

RE4DW is a model-based goal-oriented RE framework that from a business strategy supports obtaining and structuring the data to be stored in a DW for monitoring purposes. It consists of two modelling components namely, context modelling and data modelling, and a set of steps that guides building these models.

2.1   Context Modelling

The context modelling aims to identify what organisational data is required to design a DW that supports tracking KPIs. For the context modelling, we use the i* framework [17] which is a goal-oriented approach to requirement engineering. The framework is composed of two modelling components, namely, the Strategic Dependency (SD) model and the Strategic Rationale (SR) model. An *SD model* is used to describe how various actors are dependent on one another in an organisational context. An *SR model* is captured for each actor and provides a structure for expressing the rationales behind dependencies in an SD model.

We identify the major business stakeholders such as decision-makers and business process owners, and model them as social actors in the SD model. Each of these actors might have many instances. For example, CEOs or executive managers who follow the fulfillment of business strategies are instances of decision makers and operational mangers are business owners who mange business processes to accomplish a strategy. As in [17], we model DW as an actor as well. Actors depend on one another to achieve the strategic goals. Next, we model in an SR model the details about how to accomplish the strategic goals. The SR model for the decision-maker actor aims to give a full picture of the goals that she wishes to monitor in order to track the fulfillment of the strategic goals. Each strategic goal is further decomposed into subgoals and links among the subgoals are established. For each goal (strategic goals and subgoals), we model a KPI that represents its degree of fulfillment. DW actor's SR model captures the goals, tasks, and resources of the DW required to compute KPIs captured in decision-maker's SR model. To do so, we break down KPIs into their respective data components. We remark that the use of the i* framework in the DW context requires introducing new concepts and corresponding graphical notations such as indicator (represented by $\triangle$), context of analysis (represented by φ), and measurement coordinate (represented by δ). Next, we provide a set of steps to support building the SD and SR models.

Step1: As the first step, build the SD model for all actors including decision-makers, business owners, and a DW. Next, establish the dependencies among the actors. For that, model the goal dependency and task dependency between the decision-maker and business owner actor as decision-makers depends on business owners to achieve strategy by performing business processes. Model the resource dependency between DW, business owners, and decision-makers as decision-makers and business owners depend on DW to access and analyse data to evaluate the status of strategic goals by means of KPIs.

Step 2: Build the SR model for the decision maker actor. Then, capture the strategic goal defined according to step 1. Decompose the strategic goals into the subgoals of business processes which were captured in the SD model. Next, connect these subgoals with its respective strategic goals with an *AND* decomposition link. Decompose further each subgoal by asking how and how else questions up till the level of details that the decision-maker desires to monitor. Assign indicators to each goal (strategic goals and subgoals) which quantify to which degree the corresponding goal is fulfilled.

Step 3: Build the SR model for the DW actor by representing a goal for each KPI captured in Step 2. In order to guide how to define each goal, ask "what is going to be analysed with a KPI". Call these goals analysis goal to differentiate it from business goals captured in the SR model for the decision maker actor. Map the links from the SR model for the decision maker actor exactly in the SR model for the DW actor.

Step 4: Define the measures needed to be calculated for each analysis goal defined in Step 3. Represent each measure using a task and connect it to the corresponding analysis goal with a decomposition link. Next, define the coordinates for each measure. The coordinate is a property of a measure and provides the aspects for which a measure is calculated. Represent the coordinates with small circles connected to tasks representing the measures.

Step 5: Identify the data required to compute each measure defined in Step 4. To do so, identify an event required to be recorded in order to be able to compute a measure. The event includes a collection of basic numerical and non-numerical attributes. We call these events field. Identify the numerical attribute of the field as an analysis focus which reflects the quantitative aspect of a KPI. Represent the analysis focus with a resource and connect it to the task captured in Step 4 using a means-end link. Identify the non-numerical attributes as analysis contexts which is served to supply data to coordinates identified in Step 4. Represent the analysis context with small squares connected to the respective resource.

2.2   Data Modelling

Data modelling aims to derive the MD model from the goal models of the context modelling. To assist the derivation, we use an indicator matrix and a representation of data sources as a UML class diagram. The matrix (inspired by the DW bus matrix [8]) is a communication tool to summarize the KPIs information of the goal model. A UML class diagram which provides data representation within a domain is used to provide a unified illustration of the available or desired data sources to assist the design of the DW. In a MD model, data is perceived in an n-dimensional space, called a *data cube*. A data cube is based on the notions of dimensions and facts. The cells of the data cube, called *facts*, are the data being the subject of analysis. Facts have associated with them numeric values, called *measures*, which are used to quantitatively evaluate the various aspects of data. *Dimensions* provide perspectives to analyse the facts. Dimensions consist of discrete alphanumeric attributes which are organized in *hierarchies*. Each set of distinct attributes of a dimension hierarchy is called a *level*. Instances of levels are called *members*. Hierarchies specify the order of members from the most generic at the top to the most specific level at the bottom. We use the MultiDim model [15] to graphically represent an MD model. Next, we provide the steps to derive an MD model from a goal model.

Step 6: Structure the elements attached to the tasks and the resources in rows and columns of a matrix. Group rows based on the analysis focus of a KPI represented with a resource. Add tasks identified in Step 4 as columns with the measures as well as attributes of fields identified in Step 5. Fill the cells of the matrix based on the tasks and resources with their associated properties.

Step 7: Identify the resources representing the analysis focus as a fact of the MD model. Assign a measure to the fact if there is a corresponding row in the matrix defined in Step 6.

Step 8: Explore the data sources to identify the dimensions for facts. To do so, represent a conceptual model of the data sources by means of a UML class diagram. Map the analysis focus identified in Step 5 to an attribute of a class in the class diagram. Search the same class to map the contexts identified in Step 5. Define a dimension for the fact, If a context is mapped to an attribute in the same class to which the focus of analysis is mapped.

Step 9: To obtain the dimension hierarchies, use both the matrix and the data sources. Search the matrix to find the dimension of a certain fact which is represented by an attribute of the field in the matrix. Map each attribute of the field to an attribute of the data source. Look for non-numerical attributes of the same field in the matrix that are not mapped to the class of the data sources where the dimension exists. This non-numerical attribute should be find in the class diagram of data sources. Search the classes directly linked to the class containing the dimension to map the attributes of the field. Identify a hierarchy for the dimension if the attribute of the field is mapped to an attribute of a class.

3   Case Study

We illustrate the use of RE4DW with the help of a use case from the energy sector. In Poland, a number of smart energy grids are under development to meet the energy needs of a town of 10–20k inhabitants. A grid includes a physical infrastructure and an information system. A part of the information system will be an analytical system focusing on energy production and consumption analysis. The stakeholders of the analytical system want to follow up the strategic goal related to the energy efficiency. The strategic goal is to increase the energy efficiency which means to balance the energy production and consumption while at the same time to decrease the amount of both. The objective is to design a DW that is aligned with the strategic goal and is able to support the monitoring of its fulfillment. To do so, we apply RE4DW to the presented case.

The context modeling starts with building the SD model. In the case presented, the *executive board* as the decision maker actor defines the strategic goal of *Improving the energy efficiency for each city 1.5% yearly*. To improve efficiency, they need to balance between energy production and consumption. The goal is achieved through *Energy production management* and *Energy consumption management* business processes which are run by *Production manager* and *Consumption manager* as the business owner actors, respectively. Figure 1 represents the dependency among decision-makers and business owners actors to accomplish the strategic goal. Figure 1 also represents that the social actors depends on the *DW* actors to supply them the required information to track the achievement of the business strategy.

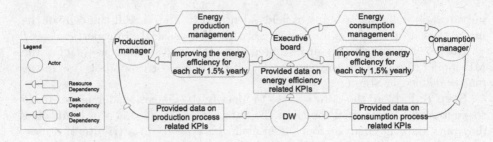

Fig. 1. Strategic dependency model incorporating the DW actor

The next step is to build the SR model for *Executive board* according to Step 2. Figure 2 represents that *G1: Improving the energy efficiency for each city 1.5% yearly* needs to balance the energy consumption and production by decreasing both. Therefore, it is decomposed into *G1.1: Decreasing the energy consumption for each city 15% yearly* and *G1.2: Increasing the production efficiency for non-performant grids 10% yearly*. Figure 2 illustrates the entire decomposition of the strategic goal of the case study. The figure also represents the indicators defined to quantify the fulfillment of the corresponding goal. For example, *KPI1: energy efficiency* quantifies to which degree *G1: Improving the energy efficiency for each city 1.5% yearly* is being achieved.

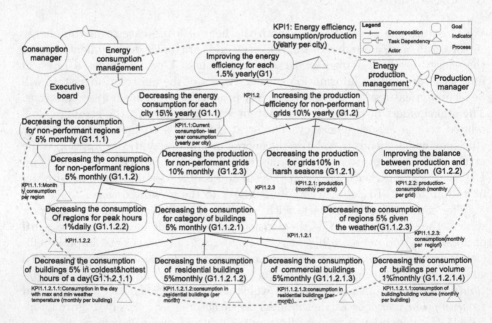

Fig. 2. Strategic rational model for decision maker actor

Next is to build the SR model for the DW actor. Figure 3 illustrates that for each KPI an analysis goal is being represented in the SR model according to Step 3. Such goal is defined by asking what is going to be analysed with, for example, *KPI1*. The answer is captured as *G-DW1: Analyzing the energy efficiency for cities yearly* which represents an intention to calculate *KPI1: consumption/production (yearly per city)*. To accomplish *G-DW1*, Fig. 3 represents that the DW actor needs *M1: Measure sum production* and *M2: Measure sum consumption*. These are the examples of a measure identified according to Step 4. Both measures are calculated for a *City* over a *Year* as the main coordinates.

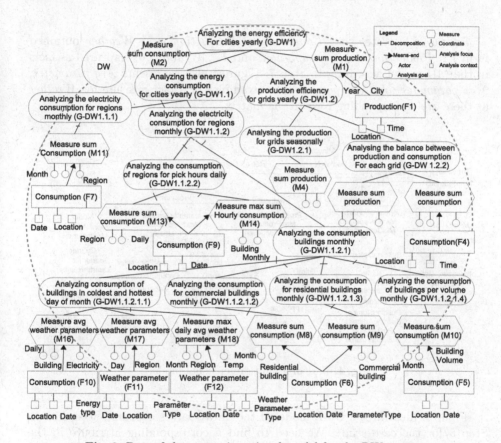

Fig. 3. Part of the strategic rational model for the DW actor

Moreover, Fig. 3 illustrates that for *M1: Measure sum production*, the DW actor needs to conduct the sum operation on a numerical value of *F1: Production* which is an example of an analysis focus identified according to Step 5. Besides, *M1* suggests to conduct the sum operation on *F1: consumption* for *city* over a *year* which are supplied by *Date* and *Location*, respectively. These are the examples of a analysis context identified according Step 5. Table 1 illustrates the template of the matrix filled with some examples from the case study.

Table 1. The template of the matrix for summarizing the data from the goal model

Analysis focus	Field Weather parameter	Date	Location	Energy type	Measure	KPI
Production		Year	City		Sum	**KPI$_1$**
Consumption		Year	City		Sum	**KPI$_1$**
		Month	Region	Electricity	Sum	**KPI$_{1.1.1}$**
Weather	Temperature	Day	Region		Avg	**KPI$_{1.1.2.1}$**
	Temperature	Month	Region		Max (Daily Avg)	**KPI$_{1.1.2.1}$**

Figure 4 illustrates that *Consumption*, *Production*, and *Weather* obtained from *F1: Production*, *F2: Consumption*, and *F3: Weather* are captured as facts in the MD model according to Step 7. Figure 4 also represents *Avg parameter*, *Max parameter* and *Min parameter* as examples of a fact measure for *Weather* as there are corresponding rows in the matrix.

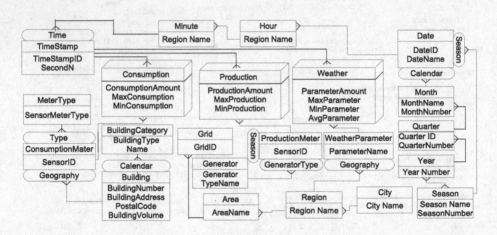

Fig. 4. Part of a multidimensional model

Figure 5 illustrates a UML model of the data sources captured according to Step 8 for the case study. We need to find a corresponding attribute in the data source for *F1: Consumption* (which we draw a fact in the MD model in Fig. 4). It is mapped to *Production Amount*, an attribute in the *Production* class of the UML model in Fig. 5. The *Production* class is searched also to map the corresponding contexts of *F1* such as: *Date* and *Location*. They are mapped to *Meter ID* and *Time Stamp*, respectively. Therefore, *Production Meter* and *Time Stamp* are the examples of the dimensions for the *Production* fact.

The MD model in Fig. 4 also shows that, *F3* includes a non-numerical property of *Generator type*. It can be mapped to the *Meter Type ID* attribute in the *Production Meter* class in Fig. 5. The class also is linked to the *Production* class

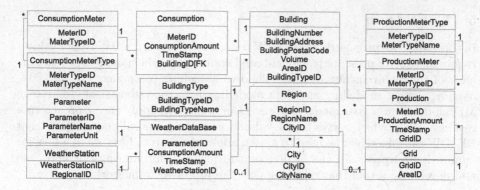

Fig. 5. Part of an UML class diagram for modeling data sources

containing an attribute which was previously mapped to the *Production meter* dimension in the MD model. Therefore, for this attribute, a dimension hierarchy is represented in the MD model.

4   Related Work

Conceptual modelling research plays a critical role in addressing challenges of designing and implementing DWs and KPIs. Research efforts in this domain have addressed a wide range of problems in business domains, from DWs (e.g., [5]) to BI (e.g., [6]). Despite their usability and expressiveness, existing conceptual modelling approaches do not address challenges of how to obtain and structure the right data from KPIs to support effectively tracking the business strategies. The existing demand-driven RE methods to design a DW use various drivers to obtain the data to be stored in DWs by which aiming to align the DWs with organisational objectives [1,4,5,9,11]. Our approach derives the data from KPIs. We aim to support and align DWs designed specifically for tracking KPIs which is the basis for analytics solution from descriptive to predictive. To the best of our knowledge, no approach took this initiative. There are also proposals in the literature [6] which include KPIs as a concept into goal models using frameworks like i* and URN, etc. These proposals offer a precise perspective over the concept of KPIs such as "what" is measured and "why" it is measured. A further approach [10] has extended goal models using i* with KPIs but the focus is on enabling business users to define KPIs more formally by using the SBVR language. Then, the OCL for OLAP [14] is introduced to validate KPIs with a DW schema. The default assumption in the above-mentioned proposals is that there is an existing DW schema and KPIs are quarried over the schema. Our RE framework took the opposite approach by deriving the data elements of a DW directly from KPIs by the means of a goal model.

5   Conclusion

In this paper, we proposed the RE4DW framework to design a DW from business requirements. We presented a model-based approach in which models provide the basis of the requirements artifacts. The two modelling components of RE4DW, namely, context modelling and data modelling help to systematically derive an MD model from a goal model using a set of steps. We apply the approach to a real case study which serves as the validation of the proposal and suggests that RE4DW can effectively be used to reason about the data to be store in a DW for monitoring purposes. In the future, we will focus on formalizing our approach and developing a tool that could support the proposed framework.

References

1. Bonifati, A., Cattaneo, F.: Designing data marts for data warehouses. ACM Trans. Softw. Eng. Methodol. **10**(4), 452–483 (2001)
2. Chowdhary, P., Mihaila, G., Lei, H.: Model driven data warehousing for business performance management. In: Proceedings of the IEEE International Conference on e-Business, Engineering, pp. 483–487 (2006)
3. Frendi, M., Salinesi, C.: Requirements engineering for data warehousing. In Proceedings of Workshop on RE: Foundation for Software Quality, pp. 75–82 (2003)
4. Gallardo, J., Giacaman, G., Meneses, C., Marbán, Ó.: Framework for decisional business and requirements modeling in data mining projects. In: Corchado, E., Yin, H. (eds.) IDEAL 2009. LNCS, vol. 5788, pp. 268–275. Springer, Heidelberg (2009). doi:10.1007/978-3-642-04394-9_33
5. Giorgini, C., Jazayeri, M., Mandrioli, D.: GRAnD: a goal-oriented approach to requirement analysis in data warehouses. Decis. Support Syst. **45**(1), 4–21 (2008)
6. Horkoff, J., Barone, D., Jiang, L., Yu, E., Amyot, D., Borgida, A., Mylopoulos, J.: Strategic business modeling: representation and reasoning. Softw. Syst. Model. **13**(3), 1015–1041 (2014)
7. Kimball, R.: The Data Warehouse Lifecycle Toolkit: Expert Methods for Designing, Developing, and Deploying Data Warehouses. Wiley, New York (1998)
8. Kimball, R., Ross, M.: The Data Warehouse Toolkit: The Complete Guide to Dimensional Modeling. Wiley, New York (2011)
9. Malinowski, E., Zimányi, E.: Requirements specification and conceptual modeling for spatial data warehouses. In: Meersman, R., Tari, Z., Herrero, P. (eds.) OTM 2006. LNCS, vol. 4278, pp. 1616–1625. Springer, Heidelberg (2006). doi:10.1007/11915072_68
10. Maté, A., Trujillo, J., Mylopoulos, J.: Conceptualizing and specifying key performance indicators in business strategy models. In: Proceedings of the Conference on the Center for Advanced Studies on Collaborative Research, pp. 102–115. IBM Corp. (2012)
11. Mazón, J., Trujillo, J., Lechtenbörger, J.: Reconciling requirement-driven data warehouses with data sources via multidimensional normal forms. Data Knowl. Eng. **63**(3), 725–751 (2007)
12. Nasiri, A., Wrembel, R., Zimányi, E.: Model-based requirements engineering for data warehouses: From multidimensional modelling to KPI monitoring. In: Jeusfeld, M., Karlapalem, K. (eds.) Proceedings of International Workshop on Conceptual Modeling. LNCS, vol. 9382, pp. 198–209. Springer, Cham (2015). doi:10.1007/978-3-319-25747-1_20

13. Nasiri, A., Zimányi, E., Wrembel, R.: Requirements engineering for data ware-houses. In: Proceedings of the Conference on Journes francophones sur les Entrepts de Donnes et l'Analyse en ligne, EDA, pp. 49–64 (2015)
14. Pardillo, J., Mazón, J., Trujillo, J.: Extending OCL for OLAP querying on con-ceptual MD models of DWs. Inf. Sci. **180**(5), 584–601 (2010)
15. Vaisman, I., Zimányi, E.: DW Systems: Design and Implementation. Springer, Heidelberg (2014)
16. Winter, R., Strauch, B.: Information requirements engineering for data warehouse systems. In: ACM Symposium on Applied, Computing, pp. 1359–1365 (2004)
17. Yu, E.: Towards modelling and reasoning support for early-phase requirements engineering. In: Proceedings of IEEE International Conference on Requirements Engineering, pp. 226–235 (1997)

Towards Formal Strategy Analysis with Goal Models and Semantic Web Technologies

Christoph G. Schuetz[✉] and Michael Schrefl

Johannes Kepler University Linz, Altenberger Str. 69, 4040 Linz, Austria
{schuetz,schrefl}@dke.uni-linz.ac.at

Abstract. An informed strategy-making process involves strategy analysis to determine the strategic position of the company. In this paper, we investigate the formalization of different frameworks of strategy analysis in order to facilitate the tasks of strategic management. The thus formalized strategic reports can be shared among employees and combined for more holistic analyses. Semantic web technologies serve as the technological foundation, which allows for the expression of strategic questions as queries over the models as well as an integration of external data sets from the semantic web.

Keywords: Knowledge management · Strategic management · Business modeling · SWOT · PESTEL · Porter's Five Forces · iStar 2.0

1 Introduction

Strategy analysis contributes towards an informed strategy-making process that ultimately secures the long-term viability of the company. Frameworks for strategy analysis serve for the identification and classification of the relevant factors for strategic decisions [5, p. 27], fostering an understanding of the company's strategic position that considers both external factors and the capabilities of the company [9, p. 11].

Goal models introduce a strategic element to conceptual modeling. As such, goal models are typically employed in early-phase requirements engineering and focus on *why* a planned system should behave in one way or the other [18]. The iStar 2.0 (i*) modeling language [4], for example, allows for capturing the actors that ultimately interact with the planned system as well as the goals and tasks that drive these interactions.

In this paper, we exemplify how goal models may be employed to formalize traditional frameworks of strategy analysis, e.g., SWOT, PESTEL, and Porter's Five Forces. The structured representation of strategic reports through goal models facilitates knowledge-sharing among employees and allows for a combination of strategic reports, e.g., from different departments over different time periods, to obtain more holistic analyses. Semantic web technologies provide the technological means for formalizing strategy analysis with goal models. Key technologies are the Resource Description Framework (RDF), RDF Schema (RDFS),

© Springer International Publishing AG 2017
S. de Cesare and U. Frank (Eds.): ER 2017 Workshops, LNCS 10651, pp. 144–153, 2017.
https://doi.org/10.1007/978-3-319-70625-2_14

the Web Ontology Language (OWL), and the SPARQL query language (see [6] for further information). Individual strategic questions – e.g., *What are the opportunities and threats for a particular company?* – can then be expressed as SPARQL queries and executed on different data sets. Furthermore, linked open data (LOD) published on the semantic web [15] could complement the knowledge about business situations.

The remainder of this paper is organized as follows. First, we briefly introduce strategy analysis and review related work. In Sect. 2, we then focus on representation of strategic reports using goal models. In Sect. 3, we focus on the use of semantic web technologies for further data analysis. We conclude with a discussion and an outlook on future work.

1.1   Strategy Analysis

Strategy analysis considers, on the one hand, the environment of the company and, on the other hand, the company itself [5]. Analysis of the environment typically consists of a study of the macro-environment followed by industry analysis [9]. Analysis of the company assesses the company's capabilities, a key instrument being value chain analysis (cf. [12]). The results of the analysis of both environment and company allow decision makers to determine the company's strategic position which, in turn, serves as an important input to making informed, rational strategic choices [9]. In this paper, we exemplify the concept of formal strategy analysis by focusing on frameworks for environmental analysis.

SWOT (strengths, weaknesses, opportunities, threats) analysis, as one of the basic analytical frameworks in strategic management, reflects the dual focus of strategy analysis on both internal and external factors of a company. Strengths and weaknesses capture the capabilities of the company. Opportunities and threats depend on the environment. Although criticized (see [5, p. 12f.]), SWOT remains a staple in the strategist's toolbox.

Apart from SWOT, more refined frameworks exist to analyze a company's environment. The PESTEL framework [9] breaks up macro-environment into political, economic, social, technological, ecological, and legal environment. For each PESTEL dimension, the analyst identifies a set of factors that determine a company's strategic position. Furthermore, Porter [13] famously identified five forces that characterize an industry: threat of substitutes, threat of new entrants, rivalry among existing competitors, as well as bargaining powers of suppliers and buyers. Both five-forces analysis and PESTEL can be part of an assessment of market attractiveness [8, p. 280].

1.2   Related Work

Related work [2,3] has investigated the representation of enterprise models using business model ontologies and semantic web technologies, with a focus on *enterprise model analysis*, i.e., checking validity against reference models and algorithmically analyzing the complexity of enterprise models. Similar to our work, the semantic web technologies RDF, RDFS, and OWL are used for the representation of business model ontologies. We, however, focus on strategic questions rather than constraint and validity checking – important tasks that could be adapted in order to construct a modeling tool and analysis client for strategy analysis.

The strategic business model ontology (SBMO) [14] adapts the i* framework for modeling a company's strategy, thereby focusing on goals, motivation, and intentions of the actors. Samavi et al. [14] position the SBMO as a methodology for requirements engineering. The rationale behind SBMO is obvious: A better understanding of the goals and intentions behind strategic actions will ultimately lead to better service design. As opposed to SBMO, we formalize strategy analysis in order to support strategy analysts rather than requirements engineers.

The business intelligence model (BIM) [7] aims at rendering business intelligence (BI) more accessible to average business users. Horkoff et al. [7] argue that current BI systems focus too much on the data in order for business users to effectively work with these systems. Business users hence expect the data models to be presented in familiar (business) terms, e.g., strategy, business models, business processes, risks. In this regard, BIM offers consolidation of the predominantly data-centric view in today's BI and the business-centric view of analysts. To this end, BIM represents goals, situations, and business processes besides the more data-oriented indicators. With our paper, we aim to position formalized strategy analysis based on goal models and semantic web technologies as a means to knowledge management in strategic management. We aim to make explicit the various strategic reports currently otherwise compiled in companies in the form of natural-language text. The thus formalized written reports can be more easily shared and combined across individuals and departments.

2   Goal Modeling for Strategy Analysis

The PESTEL dimensions of factors, e.g., political and ecological, for analysis of a company's macro-environment translate into i* actors with an actor boundary. Each of these actors represents an abstract real-world actor to be reckoned with. For example, the political dimension translates into an actor *Politics*, representing politics at large as a force of influence in the real world. Elements within the boundary of these actors represent the specific factors in the respective PESTEL dimensions. The actors that represent PESTEL dimensions may be refined via *participates-in* relationship into several other actors which represent relevant, more concrete real-world actors in the respective dimensions. Other actors represent more concrete real-world actors affected by the PESTEL factors, e.g., individual companies or types of companies. Elements within the boundaries of these other actors usually depend on the PESTEL actors' elements.

Figure 1 shows political and ecological factors of a PESTEL analysis of the airline industry's macro-environment. The example follows a case study of the low-fare airline Ryanair's strategic position [9] and is partially based on a PESTEL analysis of the airline industry [8, p. 56]. The political dimension, represented by the *Politics* actor, comprises national governments and the European

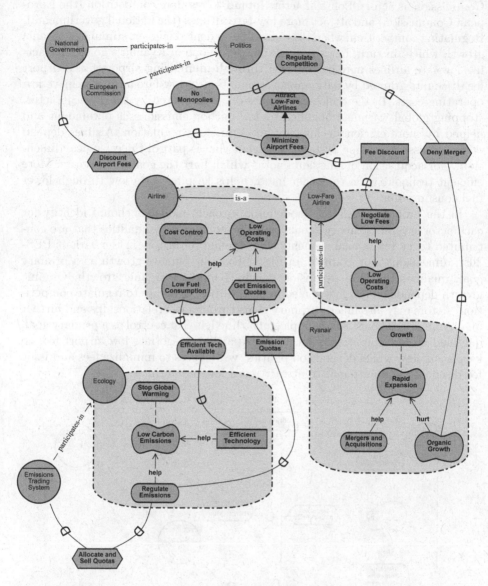

Fig. 1. Political and ecological factors of a PESTEL analysis of the airline industry's macro-environment, formalized using i* (case adapted from [9, p. 612ff.], model edited with piStar [11]). A circle denotes an actor, a dash-dotted circle attached to an actor denotes the actor's boundary. A circle with a horizontal line in the top half denotes an agent, i.e., a concrete actor. An ellipse denotes a goal, a hexagon denotes a task, a rectangle denotes a resource, a potato-shaped form denotes a quality which may be linked to a goal, task or resource by a dashed line. A connecting line with the letter "D" denotes a dependency. The direction of the "D" indicates the dependency's direction.

Commission as actors. Political actors intend to regulate competition (the European Commission) and attract more low-fare airlines (the national governments). Regulating competition, when the outcome is a denied merger, stimulates organic growth which, in turn, hurts rapid expansion – one of Ryanair's goals. Attracting low-fare airlines may be achieved through minimizing airport fees. Airport fee discounts granted by national governments help low-fare airlines achieve low operating costs. In the ecological dimension, represented by the *Ecology* actor, stopping global warming, qualified by low carbon emissions, is paramount and helped by more efficient technology and emissions regulation. Airlines depend on the Emission Trading System – an agent that is part of the ecological dimension – for acquisition of emission quotas which hurt the goal of low costs. More efficient technology, on the other hand, helps keeping costs low through lower fuel consumption.

In the proposed PESTEL modeling approach, modelers should identify for each actor several "primary" goals which are qualified by qualities that are contributed to by intentional elements that depend on factors in the various PESTEL dimensions. For example, in Fig. 1, Ryanair pursues growth as a primary goal, qualified as rapid expansion which is hurt by organic growth. Organic growth depends on mergers denied under politics' mandate to regulate competition. Actors may also inherit primary goals through is-a relationships and further qualify inherited goals. For example, an airline has cost control as a primary goal, qualified by low operating costs. Low-fare airlines negotiate low airport fees to keep costs low, which depends on politics' willingness to minimize fees and issue fee discounts in an attempt to attract more low-fare airlines.

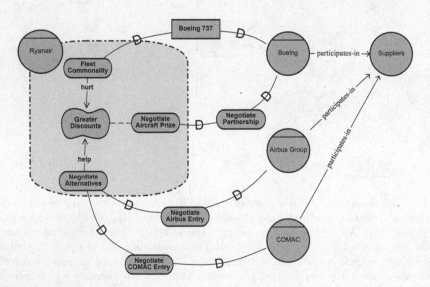

Fig. 2. Part of an industry analysis focusing on Ryanair's suppliers (case adapted from [9, p. 615f.])

An analysis of Porter's Five Forces requires identification of suppliers and customers, possible future competitors and potential substitutes. Similar to the PESTEL representation, each force translates into an i* actor representing the respective force, e.g., suppliers, as an abstract real-world actor which other actors participate in, representing more concrete real-world actors, e.g., Boeing, Airbus Group, and COMAC. Figure 2 illustrates the dependencies between Ryanair and its (prospective) suppliers modeled using i*. Ryanair commands a homogeneous fleet of Boeing 737 aircraft. The commitment to a single type of aircraft hurts the target of obtaining greater discounts during aircraft prize negotiations, since Ryanair's fleet commonality policy is publicly known. Establishing a credible threat of alternative suppliers, however, will increase Ryanair's leverage over Boeing to negotiate greater discounts.

3 Semantic Web Technologies for Data Analysis

We employ semantic web technologies to formalize strategic reports for use in knowledge-based systems. Figure 3 proposes an RDF representation for i*. The schema follows the iStar 2.0 metamodel [4, p. 14] but refrains from using reification in order to keep the graph structure of the RDF representation close to the visual representation. For example, depends-on is a property from an actor or intentional element to an actor or intentional element rather than a class as in the iStar 2.0 metamodel; domain and range of depends-on can only be represented in OWL (not shown). The RDF representation also introduces the related-with property as an abstraction of is-a and participates-in (Fig. 3, Lines 5 and 6). The absence of reification facilitates query formulation. Listing 1 then shows an RDF representation of the political dimension of the PESTEL analysis in Fig. 1; classes and properties that are defined in Fig. 3 have an istar prefix.

$Agent \sqsubseteq Actor$	(1)	$Task \sqsubseteq IntentionalElement$	(12)
$Role \sqsubseteq Actor$	(2)	$Resource \sqsubseteq IntentionalElement$	(13)
$\exists related\text{-}with.\top \sqsubseteq Actor$	(3)	$\exists contributes\text{-}to.\top \sqsubseteq IntentionalElement$	(14)
$\top \sqsubseteq \forall related\text{-}with.Actor$	(4)	$\top \sqsubseteq \forall contributes\text{-}to.Quality$	(15)
$is\text{-}a \sqsubseteq related\text{-}with$	(5)	$helps \sqsubseteq contributes\text{-}to$	(16)
$participates\text{-}in \sqsubseteq related\text{-}with$	(6)	$hurts \sqsubseteq contributes\text{-}to$	(17)
$\exists wants.\top \sqsubseteq Actor$	(7)	$\exists needed\text{-}by.\top \sqsubseteq Resource$	(18)
$\top \sqsubseteq \forall wants.IntentionalElement$	(8)	$\top \sqsubseteq \forall needed\text{-}by.Task$	(19)
$Property(depends\text{-}on)$	(9)	$\exists qualifies.\top \sqsubseteq Quality$	(20)
$Goal \sqsubseteq IntentionalElement$	(10)	$refines\text{-}and \sqsubseteq refines$	(21)
$Quality \sqsubseteq IntentionalElement$	(11)	$refines\text{-}or \sqsubseteq refines$	(22)

Fig. 3. Description-logic vocabulary for i*, expressible in RDFS

Listing 1. RDF representation of political dimension of PESTEL analysis in Fig. 1

```
1  :Politics istar:wants :Regulate_Competition ,
2    :Attract_Low-Fare_Airlines , :Minimize_Airport_Fees .
3  :National_Government istar:particpates-in :Politics .
4  :European_Commission rdf:type istar:Agent ;
5    istar:particpates-in :Politics .
6  :Regulate_Competition rdf:type istar:Goal ;
7    istar:depends-on :No_Monopolies.
8  :No_Monopolies rdf:type istar:Quality ;
9    istar:depends-on :European_Commission .
10 :Attract_Low-Fare_Airlines rdf:type istar:Goal .
11 :Minimize_Airport_Fees rdf:type istar:Goal ;
12   istar:depends-on :Discount_Airport_Fees ;
13   istar:refines-or :Attract_Low-Fare_Airlines .
14 :Discount_Airport_Fees rdf:type istar:Task ;
15   istar:depends-on :National_Government .
16 :Airline istar:wants :Cost_Control ,
17   :Low_Operating_Costs .
18 :Low_Operating_Costs istar:qualifies :Cost_Control .
19 :Low-Fare_Airline istar:is-a :Airline ;
20   istar:wants :Negotiate_Low_Fees ,
21     :Low_Operating_Costs .
22 :Negotiate_Low_Fees rdf:type istar:Goal ;
23   istar:depends-on :Fee_Discount ;
24   istar:helps :Low_Operating_Costs .
25 :Fee_Discount istar:depends-on :Minimize_Airport_Fees .
26 :Ryanair istar:participates-in :Low-Fare_Airline ;
27   istar:wants :Growth , :Rapid_Expansion ,
28     :Mergers_and_Acquisitions , :Organic_Growth .
29 :Growth rdf:type istar:Goal .
30 :Rapid_Expansion istar:qualifies :Growth .
31 :Mergers_and_Acquisitions rdf:type istar:Goal ;
32   istar:helps :Rapid_Expansion .
33 :Organic_Growth rdf:type :Quality ;
34   istar:depends-on :Deny_Merger ;
35   istar:hurts :Rapid_Expansion .
36 :Deny_Merger rdf:type istar:Task ;
37   istar:depends-on :Regulate_Competition .
```

We can classify PESTEL factors as opportunities or threats using a SPARQL SELECT query (Listing 2) over the corresponding RDF representation. Each result tuple of the query classifies a factor as opportunity or threat for a particular actor. The classifications are expressed using the classes Opportunity and Threat (with a swot prefix). Hence, a PESTEL factor is classified an opportunity or threat for an actor if one of that actor's primary goals is qualified by a quality that is helped or hurt, respectively, by an intentional element that (transitively) depends on the

Listing 2. Generic SPARQL query to classify factors as opportunities or threats

```
1  SELECT DISTINCT ?actor ?factor ?classification WHERE {
2    ?actor istar:related-with*/istar:wants ?goal .
3    ?quality istar:qualifies ?goal .
4    {
5      ?actor istar:related-with*/istar:wants ?help .
6      ?help istar:helps+ ?quality .
7      ?help istar:depends-on+/istar:refines* ?factor .
8      ?dimension istar:wants ?factor .
9      BIND(swot:Opportunity AS ?classification)
10   } UNION {
11     ?actor istar:related-with*/istar:wants ?hurt .
12     ?hurt istar:hurts+ ?quality .
13     ?hurt istar:depends-on+/istar:refines* ?factor .
14     ?dimension istar:wants ?factor .
15     BIND(swot:Threat AS ?classification)
16   }
17 }
```

PESTEL factor in question. Consider, for example, the RDF data set in Listing 1. The goals Minimize_Airport_Fees and Attract_Low-Fare_Airlines would be classified opportunities for low-fare airlines, the goal Regulate_Competition a threat for Ryanair[1]. Regulate_Competition becomes a threat for Ryanair via the dependency of Organic_Growth which hurts Rapid_Expansion, a qualifier of the primary goal Growth. Note that the SPARQL query in Listing 2 requires RDFS reasoning to be performed prior to query execution.

Other SPARQL queries, possibly in combination with external data sources, could also serve to formalize industry analysis using Porter's Five Forces. For example, in Fig. 2, the bargaining power of Ryanair's suppliers could be determined by counting the number of suppliers in a relationship with Ryanair. With a more comprehensive model, more complex graph analysis could also serve for computing the characteristics of Porter's Five Forces. Furthermore, using SPARQL's SERVICE clause, external data sources such as DBPedia[2] and wikidata[3] could be integrated into the analysis, e.g., to compute the bargaining power of suppliers and customers, or automatically determine potential suppliers, customers, and substitutes. In particular, these external sources could provide company facts such as revenue and number of employees. In that case, the resources in the analysis would have to be linked to the external data sources via OWL's sameAs property or similar.

---

[1] This threat is a reference to Ryanair's attempted takeover of Aer Lingus starting in 2007, which was eventually blocked by the EU Commission [9, p. 617ff.].

[2] http://wiki.dbpedia.org/.

[3] https://www.wikidata.org/.

4   Discussion and Future Work

Although intuition and creativity certainly are key drivers of successful strategizing [10], analytical and rational approaches to strategy-making are important complements for spontaneous action (see [5, p. 26] for more information on that discussion). In this sense, the formalization of strategy analysis using goal models and semantic web technologies must be regarded as complementary to the human element in strategizing, a form of knowledge management.

Future work will investigate the required organizational reengineering efforts as well as the associated technological aspects: Organizations must put in place a system to acquire, formalize and use strategic knowledge. Concerning knowledge elicitation, we assume that strategy reports are often already available in textual form, compiled by strategic managers; these written reports must then be (semi-automatically) translated into ontologies. Furthermore, future work will investigate alignment with common methodologies and frameworks in knowledge management, e.g., the CommonKADS methodology [16].

Strategic reports formalized using semantic web technologies may be organized in OLAP cubes with ontology-valued measures [17]. The dimensions of such a cube set the context for the knowledge, serialized in RDF format, codified in the measures. The measures focus on complex dependencies between entities rather than condensing complex data and knowledge into a single numeric indicator. The dimensions typically represent provenance information, e.g., the department that compiled the report, and meta-information such as the timespan covered by the strategy report or the employed modeling language. Analysts may choose and combine strategy reports using the dimensions. The combined knowledge can be further analyzed using dedicated query operators. An OLAP system with ontology-valued measures then becomes a valuable tool for managing a company's strategic knowledge. Otherwise, in the "as-is" scenario, strategic reports have to be stored in textual form, possibly in different layouts, and with different writing styles and text structures, thus hampering combination of the knowledge in the reports through analysts. Also, common analytical questions cannot be expressed unambiguously and in a reusable form as with SPARQL queries.

Since strategy analysts are typically not IT experts familiar with semantic web technologies, future work will develop graphical modeling tools with integrated support for data analysis. The graphical notation will be based on i*, possibly adapting the syntax rules to facilitate modeling for strategy analysis. Translation of the graphical model into RDFS allows for a SPARQL-based implementation of data analysis. The MetaEdit+ domain-specific modeling environment[4] may serve to implement a modeling tool. In order to evaluate the approach, future work will conduct usability studies with domain experts in strategic management. Furthermore, depending on the employed framework for strategy analysis, business model ontologies such as e³value and REA (see [1]) may be more suitable to represent strategic analyses. Knowledge required for one analysis framework could also be derived from knowledge modeled in another framework.

---

[4] http://www.metacase.com/products.html.

References

1. Andersson, B., et al.: Towards a reference ontology for business models. In: Embley, D.W., Olivé, A., Ram, S. (eds.) ER 2006. LNCS, vol. 4215, pp. 482–496. Springer, Heidelberg (2006). doi:10.1007/11901181_36
2. Caetano, A., Antunes, G., Bakhshandeh, M., Borbinha, J., da Silva, M.M.: Analysis of federated business models: an application to the business model canvas, ArchiMate, and e3value. In: 17th IEEE Conference on Business Informatics (2015)
3. Caetano, A., Antunes, G., Pombinho, J., Bakhshandeh, M., Granjo, J., Borbinha, J., da Silva, M.M.: Representation and analysis of enterprise models with semantic techniques: an application to ArchiMate, e3value and business model canvas. Knowl. Inf. Syst. **50**(1), 315–346 (2017)
4. Dalpiaz, F., Franch, X., Horkoff, J.: iStar 2.0 language guide - version 3. CoRR abs/1605.07767v3 (2016). https://arxiv.org/abs/1605.07767v3
5. Grant, R.M.: Contemporary Strategy Analysis, 9th edn. Wiley, New York (2010)
6. Hitzler, P., Krotzsch, M., Rudolph, S.: Foundations of Semantic Web Technologies. CRC Press, Boca Raton (2009)
7. Horkoff, J., Barone, D., Jiang, L., Yu, E., Amyot, D., Borgida, A., Mylopoulos, J.: Strategic business modeling: representation and reasoning. Softw. Syst. Model. **13**(3), 1015–1041 (2014)
8. Johnson, G., Scholes, K., Whittington, R.: Exploring Corporate Strategy, 8th edn. Pearson, London (2008)
9. Johnson, G., Whittington, R., Scholes, K., Angwin, D., Regnér, P., Pyle, S.: Exploring Strategy: Text and Cases, 10th edn. Pearson, London (2014)
10. Mintzberg, H.: Crafting Strategy, vol. 65. Harvard Business Review, Boston (1987)
11. Pimentel, J.: piStar Tool - Goal Modeling. http://www.cin.ufpe.br/jhcp/pistar/. Accessed 7 Aug 2017
12. Porter, M.E.: Competitive Advantage: Creating and Sustaining Superior Performance. Free Press, New York (1985)
13. Porter, M.E.: Competitive Strategy: Techniques for Analyzing Industries and Competitors, 12th edn. Free Press, New York (2013)
14. Samavi, R., Yu, E.S.K., Topaloglou, T.: Strategic reasoning about business models: a conceptual modeling approach. Inf. Syst. e-Bus. Manag. **7**(2), 171–198 (2009)
15. Schmachtenberg, M., Bizer, C., Paulheim, H.: Adoption of the linked data best practices in different topical domains. In: Mika, P., et al. (eds.) ISWC 2014. LNCS, vol. 8796, pp. 245–260. Springer, Cham (2014). doi:10.1007/978-3-319-11964-9_16
16. Schreiber, G.: Knowledge Engineering and Management: The CommonKADS Methodology. MIT press, Cambridge (2000)
17. Schütz, C., Neumayr, B., Schrefl, M.: Business model ontologies in OLAP cubes. In: Salinesi, C., Norrie, M.C., Pastor, Ó. (eds.) CAiSE 2013. LNCS, vol. 7908, pp. 514–529. Springer, Heidelberg (2013). doi:10.1007/978-3-642-38709-8_33
18. Yu, E.S.K.: Towards modelling and reasoning support for early-phase requirements engineering. In: Proceedings of the 3rd IEEE International Symposium on Requirements Engineering, pp. 226–235 (1997)

Assisting Process Modeling by Identifying Business Process Elements in Natural Language Texts

Renato César Borges Ferreira[1], Lucinéia Heloisa Thom[1(✉)],
José Palazzo Moreira de Oliveira[1], Diego Toralles Avila[1],
Rubens Ideron dos Santos[1], and Marcelo Fantinato[2]

[1] Institute of Informatics, Federal University of Rio Grande do Sul,
Av. Bento Gonçalves, Porto Alegre 9500, Brazil
{renato.borges,lucineia,palazzo,dtavila,risantos}@inf.ufrgs.br
[2] School of Arts, Sciences and Humanities, University of São Paulo,
Rua Arlindo Bettio, São Paulo 1000, Brazil
m.fantinato@usp.br

Abstract. Process modeling plays a significant role in the business process lifecycle, as it must stress the quality of process models for supporting all the next steps. However, this phase is time consuming and expensive, a consequence of the huge amount of unstructured input information. In a previous research, we presented an approach for identifying business process elements in natural language texts which facilitate the modeler's work. Such approach relies on a set of mapping rules associated with natural language processing techniques. The identification itself was already validated, but how to apply this information to minimize the modelers' effort remains unclear. Highlighting the identified rules in the text can enhance its comprehensibility. This paper explores the applicability of such mapping rules on supporting the modeler by marked up texts. The validation shows promising results, as the time spent and effort perceived by the modeler were both minimized.

Keywords: Process models · Natural language processing · Process element · Business process management · Business process model and notation · Process modeling

1 Introduction

As a phase of the business process life cycle, process modeling is very important and complex. It consists of elaborating a comprehensive description of the process as performed in an organization. This is done to facilitate information sharing and enhance understandability [1]. As the business actors involved in the process are not qualified to create formal models [2], this task is usually done by a Business Process Management (BPM) analyst, a specialist on formal documentation and aiming to improve business processes in organizations.

© Springer International Publishing AG 2017
S. de Cesare and U. Frank (Eds.): ER 2017 Workshops, LNCS 10651, pp. 154–163, 2017.
https://doi.org/10.1007/978-3-319-70625-2_15

There are several methods by which the BPM specialists may acquire information about a process during the modeling stage, such as interviews, workshops and analysis of the existing text documents (e.g.: e-mails, event logs, manuals, web pages, forms, reports). As evidences show, 85% of the information in organizations is stored as text documents [3] and this paperwork is getting bigger as information stored in texts are growing faster than conventional structured data [4]. Those documents are often dispersed within different departments, making the work even harder. In addition, most documents contain specific domain words, so that beginner modelers encounter difficulties in understanding it, suggesting that only skilled process analysts are able to perform good process modeling from texts [5–7].

As manual modeling becomes unfeasible due to scale of data collection, the need of automation support arises. The automatic identification of process elements in descriptive document texts may facilitate the work done by the specialist. However, most documents are not prepared to be directly used by common automated modeling tools, as they are fully unstructured texts [1]. Hence, current research seeks for approaches to identify business process elements using natural language processing techniques [8]. One promising way of doing this is by pre-processing the natural language text [9].

In our previous work [10], we proposed a semi-automatic method to identify business process elements from natural language texts in order to minimize the effort of BPM specialists in the process modeling and, also, to take the first step towards generating process oriented text. This method used mapping rules between syntactic relations and process elements to indicate which sentences could contain types of process model elements. The approach was validated in two ways: (i) a comparison of the results of a prototype to those of expert modelers and, (ii) a survey which verified if process modelers use the mapping rules correctly and presented them with a sentence and a process model, to check if they believed that said process model represented that sentence. This work showed promising results about the correctness of the rules, however their effectiveness on minimizing the specialist effort on extracting process elements remained not proved.

The objective of the present work is to develop a test of the applicability of the mapping rules to minimize the effort during process modeling. To perform this evaluation the approach was further validated through the application of an evaluation to measure the time necessary to model with and without a rule-mapped text and to gather feedback about the acceptance of these rules.

This paper is structured as follows. Section 2 presents the related works. Section 3 reviews the approach presented in the previous work. Section 4 shows the evaluation method and results analysis. Finally, Sect. 5 concludes the paper.

2   Related Works

The state of the art related to natural language processing and BPM can be divided in two categories: (i) the extraction of process models from natural language text and, (ii) text generation from process models.

The first category considers works that propose to build process models automatically by analyzing texts that describe a process. In the work of Friedrich et al. [8] the method first parses and analyses the syntax of the text and extracts the actors and actions from the sentences. After this, the semantic analysis is performed to find indicators for conditions, parallelism, sequence and exceptions and to determine the relationship between the sentences; also, an anaphora resolution is applied to identify the references made by determiners and pronouns. Finally, a process model is generated based on the extracted data. A limitation of this work arrives from the necessity of the text to be grammatically correct for the English language, meaning that it is necessary to perform a good grammatical correction before the analyses. Additionally, the text must not contain interrogative sentences (e.g., a sentence that ends with a question mark) and needs to be described sequentially. We conducted an introductory approach to solve this problem [9], in which it was concluded that a natural language text must be processed before the extraction of process models.

Another kind of approach in this category focuses on creating process models from natural language texts using techniques usually associated with text mining and natural language interpretation. For example, [11,12] explore the use of narrative techniques for extracting process models from group stories. Each story relates the point of view of one teller and multiple stories may relate to a single general process model. This probably causes ambiguities as different tellers may tell contradictory information in their stories.

The second category, text generation from process models, has the opposite objective [5–7]. It seeks to verbalize a process model into a natural language text that describes the process. This alternative should improve the understanding of the process for business professionals that have difficulty in understanding the semantics of the process model's notation. With this improvement, these business professionals will be able to provide a better validation of the semantics of the process models for which they are responsible.

The existence of multiple documents representing a process in multiple different formats can represent an additional problem. If a descriptive text is developed and maintained independently of the represented process model, inconsistencies certainly will happen, such as ambiguous or contradicting information. Van der Aalst et al. [13,14] try to address this problem, since it is vital for the understanding of a process that the describing documents are consistent with the model.

3   An Approach to Identify Process Elements in Natural Language Texts

In [10], we proposed a method of identifying business process elements in natural language texts. As shown in Fig. 1, the method is composed of four steps executed sequentially.

The first challenge (see "Step 1. Input Data") on automating this identification is the complexity of natural languages. There are several classifications

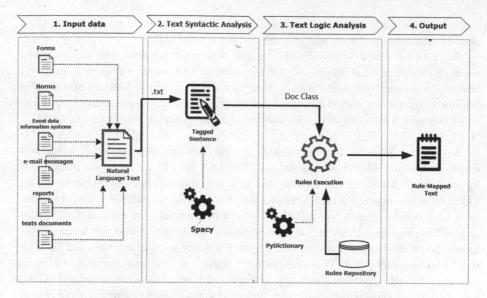

Fig. 1. A semi-automatic approach for to identify process elements in natural language texts

of natural language texts, each one with different characteristics. One particular classification is that of sequential (or procedural) texts, that are organized by stages or, in the case of describing a process, in the order it's executed. In addition, specific words (e.g., first, second, next, then, finally) occur frequently in sequential texts and may be correlated to business process elements. Alternatively, Jiexun et al. [15] claim that extraction of model fragments can be executed on heterogeneous texts as well.

To identify these fragments in a text, each sentence needs to be syntactically analyzed by a parser to produce tagged sentences which contains all the essential syntactic relations (see "Step 2. Text Syntactic Analysis"). This allows us to determine the syntactic structure of the sentence, focusing primarily on the part of speech tagging (e.g., verb, adverb, adjective, subject, direct object, indirect object etc.) and the dependency relations, as they are vital to identify process elements.

To do this, we use Spacy[1], an open-source library for natural language processing on Python. It was chosen due to its overall accuracy (90,53% average) [16], due to the its good execution time among its equivalents (according to our tests) and due to its support for all the requirements for the development of our prototype. Spacy takes sentences that have been manually separated from a text, generates their syntactic trees containing the morphological classes and, based on this tree, produces a set of tagged sentences. Each sentence is represented in a DOC class, which is a vector containing an object for each word of the sentence. Each object has the information about all the features of the word

---

[1] https://spacy.io/.

(e.g., its tokenization, sentence recognition, part of speech tagging, lemmatization, dependency parsing, and named entity recognition).

After the syntactic analysis, the tagged sentences are analyzed according to a set of mapping rules and word correlations, so that each sentence is mapped into a process element (see "Step 3. Text Logic Analysis"). These mapping rules originate from a diversified set of grammatical classes and dependency relations. They were defined manually after the search for patterns on example texts in [1,17].

In order to define such rules, a set of process elements was needed. We selected Business Process Model and Notation (BPMN), an official notation supported by the Object Management Group [18]. It is a graphical representation for specifying business processes in a business process model. As BPMN 2.0 is the most used and stable version, we relied on its set of elements for mapping the rules in process elements. However, as the complete set is very large, with many elements being used sparingly due to their very specific semantics, we used just the most common elements. This reduced set comprises activities, events, exclusive gateways (XOR), parallel gateways (AND) and lanes. Inclusive gateways (OR), despite having significant use, were excluded since they present structural and semantic problems [19–21], including semantic ambiguities [22].

The labels of potential process elements are a crucial factor for their identification. For instance, pragmatic guidelines on labeling activities Mendling [23] say that they should be composed of a verb and an object, describing the action taken and the business object of that action. Based on this, we defined the rules for activities as expecting that the sentence should contain at least a verb and an object. For example, in the sentence *"The customer sends the request"*, the label of the activity is *"Send the request"*.

This same logic is applied for events. Their label is recommended to be composed of an object followed by a participle verb (e.g., *"Invoice created"*). Therefore, the sentences that may describe an event should also contain an object and a participle verb. Current research [5,23,24] further establishes this, saying that tenses in the present and/or future denote activities, while tenses in the past and/or present perfect are denote events.

Besides verbs and objects, sentences usually contain a subject and may contain an indirect object. Both of them can refer to the lanes of a pool. In the previous example for activities, the subject of the sentence is *"The customer"*; therefore, *"Customer"* can be defined as a lane which contains the activity *"Send the Request"*.

For the identification of both XOR and AND gateways, the rules use signal words that refer to the control flow (e.g. "if", "else", "while", "in parallel"; Other words are available at https://goo.gl/p75j8F). These words describe a condition that must be checked before deciding what further actions must be done (for exclusive gateways) or the parallelism of the proceeding actions (for parallel gateways). In addition to this, we integrated a search for synonyms of these words using PyDictionary[2], a Python module that allows the synonyms retrieval from the *thesaurus*[3] database.

The complete set contains 32 rules: nine for activities, nine for events, seven for exclusive gateways, four for parallel gateways, and three for lanes. These rules were divided in two categories based on the frequency they were found in the texts analyzed: Primary rules, that happened often and are presented in table, and secondary rules, that happened rarely and are omitted from this paper. The primary and secondary rules are available at https://goo.gl/p75j8F and https://goo.gl/kpdEeF, respectively.

The result of the application of these mapping rules to all sentences of a text is a rule-mapped text, comprised of the sentences and all rules that were identified on each of them (see "Step 4. Output"). Furthermore, this output can also point out whether there are possible missing elements in the text, such as missing start or end events. Ultimately, the rule-mapped text serves as a prerequisite for a preprocessed text, which follows a strict, process-oriented structure to permit the extraction of process elements from it.

4 Evaluation

In [10] the mapping rules were validated through a survey, while a developed prototype was validated by comparing its results to a manual extraction made by the researchers using the rules.

The survey collected experts' opinions on the mapping rules. After the background collection and training steps, the participants were asked to judge if the example process models correctly represented their respective descriptive sentences in a Likert [25] scale of 5°. Six sentences and models were provided to the participants, in which 2 models were modeled according to the proposed rules and the rest of them was intentionally modeled incorrectly. As shown in Fig. 2, the participants agreed with the models created according to the mapping rules. Also significant is that the participants strongly disapproved the incorrect models, showing the validity to our approach.

The prototype showed the good implementability of these rules with natural language processing tools. Comparing to the manual extraction made by the researchers according to the mapping rules, the prototype presented elevated accuracy, precision and recall averages (91.92%, 84.44%, 87.49% respectively).

Proven their efficacy and implementability, this paper aims to test the applicability of the rules on supporting the modeler via a rule-mapped text. We conducted a new survey to check the response of modelers to the aid of the mapping rules. The main objective was to evaluate the difference of modeling process when the modelers use rule-mapped text compared to the modeling done by the same modelers without this additional aid. We focused on measuring the time spent to complete the modeling task and the effort necessary as perceived by the modeler.

---

[2] https://pypi.python.org/pypi/PyDictionary.
[3] http://www.thesaurus.com.

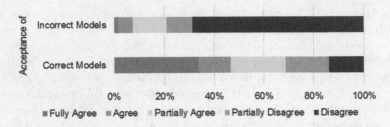

Fig. 2. Acceptance of the models created incorrectly compared to those created according to the mapping rules

The first step of the survey aimed to get the participants' background, collecting information such as profession, education, BPM and BPMN experience in years. These demographic questions permit to split the results in data groups and recognize patterns in populations.

In the second step, which collects time and difficult level of a normal process modeling, we provided a text in English describing a process model and asked the participants to model this process in BPMN. Also, we forced the participants to provide the time spent on the process modeling in minutes and the level of difficulty experienced in a Likert scale of 5 points, since this information were mandatory in order to advance to the next step. The third step was similar. We provided a different text, similar in length and complexity to the previous process, but with the addition of tags created according to the mapping rules to indicate the existence of a process model element.

Finally, the questions of the fourth step were subjective and aimed to catch the user satisfaction and experience. The participants were asked which text contained more information and if they agreed that the rule-mapped text helped in the modeling process.

The survey request was openly distributed through email, websites and social networks. In total, it had 21 participants. To catch experts' opinions, we filtered the data retrieved by the following characteristics: significant BPM and BPMN experience and more than 2 years of work as a process expert. According to [26–28], expert modelers are less likely to commit errors due to comprehension fails in models. Also, as we could not check the correctness of produced models and relied on the measure of time and effort, the filtration was done to remove the *quick and wrong* participants. This filtering reduced the number of participants to 12, representing 57.14% of the total (Fig. 3).

The average time the experts spent on modeling the rule-mapped text is 10 min, while for the first, plain, text the same modelers spent 17 min (a decrease of 41%). This shows that the rule-mapped text speeds up the modeling task. This could solve a big problem in organizations that rely on BPM, which is the time, effort and, consequently, the cost spent on the modeling phase.

Likewise, the first text presented an average of 3.25 in the difficulty scale, whereas the average of the second was 1.33, meaning that the mental effort felt by the modeler was also minimized. As mistakes occur more often in exhausted

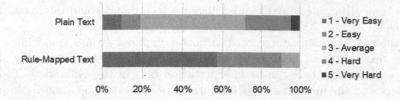

Fig. 3. Difficulty of process modeling as perceived by the participants.

modelers, this suggests that models created from rule-mapped text would be less likely to be mistaken.

Finally, the results pointed that the selected participants were unanimous on two important aspects: the rule-mapped text had more details regarding to process modeling and it minimizes the effort spent by the specialist modelers. Also, 91.66% of the participants agreed that the tags from the rule-mapped text assisted on the process modeling. Since the selected participants were specialists with modeling experience, the tags could have disturbed them during the process modeling. However, the results indicated that this hypothesis was not verified and suggest that our approach is valid. The results also show that our approach has a very high acceptance of the rule-mapped text by modelers.

5   Conclusion

The work presented in this paper shows the evaluation of the applicability of the proposed mapping rules on reducing the effort of the modeler on recognizing business process elements in natural language texts. We conducted the evaluation by comparing the process modeling from rule-mapped and plain texts through a questionnaire application. The evaluation analyzed the interaction between users and the texts in order to obtain an indication of the user satisfaction, time consumed and effort spent.

The results show not only that the mapping rules may be used to reduce effort and time on process modeling through tags but also that professional modelers highly accept the use of this method. The most impressive result is the reduction of time necessary to comprehend and model a process from a baseline model from text description (41%). This characteristic has a high impact on organizations that rely on BPM since this procedure is easily inserted on the modeling phase of the business process lifecycle. Other important result is the high acceptability of the method and the minimization of effort felt by the participants, again supporting the implementation of the method in organizations.

As limitations of this work, it must be clearly stated that it was experimented with a reduced number of rules tested. This evaluation represents 28% of the mapping rules identified in the previous research. Hence, a more comprehensive survey is needed, as the mapping rules generated do not cover the complete set of the BPMN notation. Generating the mapping rules through automated pattern recognition using artificial intelligence techniques could minimize these problems.

As the same participants worked in both compared groups (modeling rule-mapped text/modeling plain text) and the order of questions were the same for all participants, there can be interpretation bias in the evaluation. To better understand the method and its limitations, we will develop a more complete experiment, with control groups, more participants and more variables analyzed, such as semantic and pragmatic quality of the produced models. The rules were based on descriptive texts in English, different languages have inherently dissimilar structures that could relate to process model elements, which means each language needs specialized rules.

References

1. Dumas, M., La, R.M., Jan, M., Reijers, H.A.: Fundamentals of Business Process Management. Springer, Heidelberg (2013)
2. Frederiks, P.J.M., van der Weide, T.P.: Information modeling: The process and the required competencies of its participants. Data Knowl. Eng. **58**(1), 4–20 (2006)
3. Blumberg, R., Atre, S.: The problem with unstructured data. DM Review (2003)
4. White, M.: Information overlook, vol. 26, p. 7 (2003)
5. Leopold, H.: Natural Language in Business Process Models. Springer, Heidelberg (2013)
6. Meitz, M., Leopold, H., Mendling, J.: An approach to support process model validation based on text generation 33, 7–20 (2013)
7. Leopold, H., Mendling, J., Polyvyanyy, A.: Supporting process model validation through natural language generation. IEEE Trans. Softw. Eng. **40**(8), 818–840 (2014)
8. Friedrich, F., Mendling, J., Puhlmann, F.: Process model generation from natural language text. In: Mouratidis, H., Rolland, C. (eds.) CAiSE 2011. LNCS, vol. 6741, pp. 482–496. Springer, Heidelberg (2011)
9. Ferreira, R.C.B., Thom, L.H.: Uma abordagem para gerar texto orientado a processo a partir de texto em linguagem natural. In: XII Brazilian Symposium on Information Systems, vol. 77 (2016)
10. Ferreira, R.C.B., Thom, L.H., Fantinato, M.: A semi-automatic approach to identify business process elements in natural language texts. In: Proceedings of the 19th International Conference on Enterprise Information Systems. To appear (2017)
11. Santoro, F.M., Gonçalves, J.C.A., Baiao, F.A.: Business process mining from group stories. In: International Conference on Computer Supported Cooperative Work in Design, pp. 161–166 (2009)
12. Gonçalves, J.C.A., Santoro, F.M., Baiao, F.A.: Let me tell you a story - on how to build process models. J. Univ. Comput. Sci. **17**, 276–295 (2011)
13. van der Aa, H., Leopold, H., Reijers, H.A.: Detecting inconsistencies between process models and textual descriptions. In: Motahari-Nezhad, H., Recker, J., Weidlich, M. (eds.) BPM 2015. LNCS, vol. 9253, pp. 90–105. Springer, Cham (2015)
14. van der Aa, H., Leopold, H., Reijers, H.A.: Dealing with behavioral ambiguity in textual process descriptions. In: La Rosa, M., Loos, P., Pastor, O. (eds.) BPM 2016. LNCS, vol. 9850, pp. 271–288. Springer, Cham (2016)

15. Li, J., Wang, H.J., Zhang, Z., Zhao, J.L.: A policy-based process mining framework mining business policy texts for discovering process models. Inf. Syst. E-Bus. Manage. **8**(2), 169–188 (2010)
16. Choi, J.D., Palmer, M.: Guidelines for the Clear Style Constituent to Dependency Conversion. Technical Report 01–12, University of Colorado Boulder (2012)
17. Weske, M.: Business Process Management. Springer, Heidelberg (2012)
18. OMG: Business process modeling notation (bpmn). versão 2.0.2 (2013)
19. Mendling, J., Neumann, G., van der Aalst, W.: Understanding the occurrence of errors in process models based on metrics. In: Meersman, R., Tari, Z. (eds.) OTM 2007. LNCS, vol. 4803, pp. 113–130. Springer, Heidelberg (2007). doi:10.1007/978-3-540-76848-7_9
20. Figl, K., Recker, J., Mendling, J.: A study on the effects of routing symbol design on process model comprehension. Decis. Support Syst. **54**(2), 1104–1118 (2013)
21. Kossak, F., Illibauer, C., Geist, V.: Event-based gateways: open questions and inconsistencies. In: Mendling, J., Weidlich, M. (eds.) BPMN 2012. LNBIP, vol. 125, pp. 53–67. Springer, Heidelberg (2012). doi:10.1007/978-3-642-33155-8_5
22. Kindler, E.: On the semantics of epcs: Resolving the vicious circle. Data Knowl. Eng. **56**(1), 23–40 (2006)
23. Mendling, J., Reijers, H.A., van der Aalst, W.M.P.: Seven process modeling guidelines (7PMG). Inf. Softw. Technol. **52**(2), 127–136 (2010)
24. Mendling, J., Reijers, H.A., Recker, J.: Activity labeling in process modeling: Empirical insights and recommendations. Inf. Syst. **35**(4), 467–482 (2010)
25. Likert, R.: A Technique for the Measurement of Attitudes. Number N° 136–165 in A Technique for the Measurement of Attitudes. publisher not identified (1932)
26. Mendling, J., Strembeck, M., Recker, J.C.: Factors of process model comprehension: findings from a series of experiments. Decis. Support Syst. **53**(1), 195–206 (2012)
27. Mendling, J., Reijers, H.A., Cardoso, J.: What makes process models understandable? In: Alonso, G., Dadam, P., Rosemann, M. (eds.) BPM 2007. LNCS, vol. 4714, pp. 48–63. Springer, Heidelberg (2007)
28. Schrepfer, M., Wolf, J., Mendling, J., Reijers, H.A.: The impact of secondary notation on process model understanding. In: Persson, A., Stirna, J. (eds.) PoEM 2009. LNBIP, vol. 39, pp. 161–175. Springer, Heidelberg (2009). doi:10.1007/978-3-642-05352-8_13

Using Multidimensional Concepts for Detecting Problematic Sub-KPIs in Analysis Systems

Alberto Esteban[1(✉)], Alejandro Maté[2], and Juan Trujillo[1]

[1] Software and Computing Systems, University of Alicante, Alicante, Spain
{aesteban,jtrujillo}@dlsi.ua.es
[2] Lucentia Lab, Alicante, Spain
amate@lucentialab.es

Abstract. Business Intelligence, and more recently Big Data, have been steadily gaining traction in the last decade. As globalization triggers the ability for small and medium enterprises to enter worldwide markets, monitoring business objectives and pinpointing problems has become more important than ever. Previous approaches have tackled the detection of particular problematic instances (commonly called Key Performance Indicators-KPIs), trying to search for the events that are driving companies to be far away from organization's main goals. One of the key problems is that even though KPIs are positive, they are normally calculated from other sub-KPIs and therefore, it is crucial to find out which is the concrete sub-KPI that is negatively influencing the main KPI. Therefore, in this paper, we focus on a semi-automatic approach for finding the key sub-KPIs that have bad results for the company. This approach is checked on real data that are used to create a report showing potential weaknesses in order to help companies to find out which factors may affect concrete sub-KPIs. Our approach allows us to provide insights for decision makers and help them to determine the underlying problems for achieving a goal and thereby, aiding them with taking corrective actions.

Keywords: KPI · Monitoring · Business Intelligence · Reporting · Business analysis

1 Introduction

Business Intelligence and analytics have been playing an increasingly important role for monitoring enterprise activities, enabling them to take corrective actions on existing problems and guiding them to success. Most often, monitoring is based on historical and current data analysis, aided by the definition of Key Performance Indicators (KPIs) which help decision makers to focus on the most critical parts of the system. The most common way to represent KPIs is through dashboards and scorecards, presenting a comprehensive view for decision makers that link relevant data to business goals. However, the growth in data volume

© Springer International Publishing AG 2017
S. de Cesare and U. Frank (Eds.): ER 2017 Workshops, LNCS 10651, pp. 164–173, 2017.
https://doi.org/10.1007/978-3-319-70625-2_16

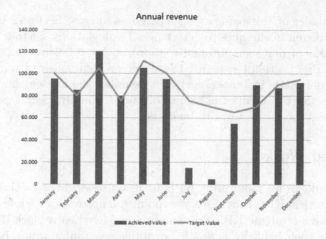

Fig. 1. Dashboard element with the annual revenue of an enterprise

has affected the ability to manage and analyze data and is introducing changes in the way that companies approach data analysis.

Indeed, an example of the current way of visualizing KPI values and evolution can be seen in Fig. 1. In this Figure we can observe the evolution in annual revenue across several months, noticing a dramatic decrease during July and August. Why is our revenue decreasing significantly during these months? Can we determine that it is a seasonal matter? Or are there other factors influencing this behaviour? As KPI values shown in dashboards are comprised of multiple dimensions contributing towards the specific slice of data being visualized (time, products, locations, activities, etc.), it becomes impossible to determine whether anomalies such as this one are perfectly normal behaviour or something that requires immediate attention from stakeholders.

This is a large problem for enterprises because we are apparently meeting our target goals, but our underlying processes are not performing as they should.

The impact of the underlying misbehavior in KPI monitoring is critical. All sectors are moving towards KPI-based operations monitored through dashboards; health, through pre-defined sets of KPIs for monitoring hospital performance, aerospace industry, where performance-based contracts (PBCs) for maintenance and components are becoming the norm, transportation, eGovernment, energy, etc. [1,2]. Therefore, advances in this area not only going to be widely applicable across all sectors but also are likely to produce an improvement in overall monitoring and performance [3–5].

Previous research has tackled this problem by mean of an algorithm to determine instances that are not accomplishing their individual targets (sub-KPI) with the purpose of taking corrective actions as soon as possible. However, nowadays, the way to identify these problematic instances is based looking at the size of the deviation of the instances with respect to their goals. Therefore, in this paper, we build on top of our previous works in order to provide insights about the problems detected.

The remainder of this paper is structured as follows. Section 2 presents an overview of different techniques to detect possible deviations. Section 3 describes the foundations of our semi-automatic approach for problem characterization within sub-KPIs. Section 4 explains the different techniques to automatically detect the problematic instances. Section 5 describes the implementation of our algorithm. Section 6 presents the results obtained. Finally, Sect. 7 presents the conclusions and the future work to be done.

2   Related Work

Existing techniques for monitoring KPIs focus on strategic or tactical KPIs [6]. Monitoring tactical KPIs provides a viewpoint too narrow for high level decision making, whereas strategic KPIs are usually considered as a block disregarding divergences in their behavior across for example geographic zones, products, or time. In order to better understand the behavior of KPIs, one of the most used techniques is SWOT Analysis [7,8].

One of the most used techniques to understand the current state of an enterprise is SWOT Analysis [7,8]. SWOT stands for Strengths (S), Weaknesses (W), Opportunities (O), and Threats (T). It contributes to the improvement of the enterprise by relating qualitative SWOT factors, detected through exploration, to strategic goals, thereby allowing the company to take corrective actions. Our approach helps to this end connecting performance-based management [9] to SWOT analysis by detecting problematic instances and potential weaknesses in enterprise KPIs, thus providing a better understanding of the enterprise context and transforming information provided by KPIs into actionable insights.

3   Previous Work

Our approach is based on multidimensional modeling where we have dimensions, hierarchies, levels and instances. Dimensions are the top-level way to represent an entity (e.g. *Product*, *Time*, etc.). Hierarchies provide structure to dimensions. Each hierarchy has some levels in order to aggregate the data. For example, we can have the *Geography* hierarchy and *Continent*, *Country*, *Region* and levels. In each level we have the instances that are the concrete elements that are contained in a level. For example, at the *Continent* level we can have the following instances: *America*, *Europe* and *Asia*.

Previously, an approach to the detection of automatic anomalies was proposed in [10]. This approach is able to detect which are the dimensions that are not achieving the main goals, looking in their hierarchies and levels and choosing the worst instances. In the following, we provide a brief explanation of the algorithm, used as a base for this work.

Initially, we have all dimensions and hierarchies. For each hierarchy of each dimension, we calculate a metric called *Information Gain* (IG) that informs how much the highest level of each hierarchy helps in isolating deviations from the target in the fewer number of instances. After the calculus, the worst instance of the level containing the highest IG is selected and added as a restriction for all future queries. Then, the pointer to the highest level of that dimension moves to the next level and another iteration begins.

The algorithm stops with a metric called *Information Lost* (IL), calculated as the size of the problem (deviation) ignored, comprised by all instances that have not been selected by the algorithm. This metric was defined in order to avoid to do an excessive search (*drilldrown*) in the multidimensional schema, with this metric we try to do not focus our attention into a very specifically instance. An example of the functionality of IL can be shown in Fig. 2.

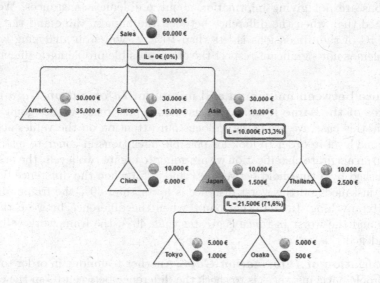

Fig. 2. Information loss example

At the beginning IL is 0, but each time a new restriction is added (e.g. *Asia*) IL is recalculated. If we define an IL threshold of 30%, the algorithm stops in *Asia*, that means that our main Sales problem is related to *Asia*. In another case, if we define an IL of 50%, the algorithm continues, looking on the next level, in that case, the restrictions obtained informs that there are problems specific to *Japan* that should be analyzed.

4 Automatic Detection of Problematic Instances

Our main goal is to extend the previous research with the aim to improve the way we stop the search. In the previous algorithm, the condition to stop searching and analyzing sub-KPIs depends in the manual definition of a threshold to IL. This means that the decision maker should have a clear idea of the size of the problem. Instead, we wish to complement this approach with the definition of rules that detect not one but different kinds of problems that are relevant to the decision maker.

Information Gain of dimensions equal to zero. The first technique that we have introduced is to stop searching for problematic instances in the case that the IG of all the dimensions that we are studying is 0. We have checked that this occurs when we have defined an excessive IL threshold.

Difference between maximum and minimum Information Gain throw all the dimensions. In order to do a more accurate halt, we are going to exploit the utility of IG. We have built-in a threshold value to check when the dimensions are not giving information about problematic instances. We have determined that when the difference between the highest value and the lowest value of IG of the dimensions is less than 10% we will stop analyzing because the problem is not significant respect the other possible problematic dimensions.

Difference between maximum and minimum Information Gain throw instances of the same level. As we have done previously we are going to use IG but in this case, we are going to focus our attention on the values achieved at the same level in order to look for possible deviations in concrete instances.

To do an accurate classification we are going to create two levels, the first level is established by a threshold between 0% and 40%, when the difference between the best instance and the worst instance is lower than 40% the framework will not notify anything. In the other hand, when the difference between the best instance and the worst instance is greater than 40%, the framework will notify this incidence.

Large variation of Information Gain. Another technique in order to try to detect problematic instances is to check the difference between IG in the current and previous iteration. We are going to focus our attention on the highest value of IG obtained in the previous iteration (in the first iteration this technique is not applicable) and compare it with the maximum IG of the current iteration.

The first level is when we obtain a difference between 0% and 25%, in this case the difference is tolerable and we do not perform any notification. The second level is when the difference is greater than 25% and lower than 50%, in this case we notify to the user that we are going to continue the search, however it is possible that real problem was in this point. Finally if the difference is greater than 50%, the algorithm stops.

Large variation of Information Lost. Now we are going to focus our attention in the IL, as we told previously, IL it is the total deviation lost as the algorithm performs restrictions. In the previous work, the way stopping algorithm was to define an IL threshold, this approach has a weak point due the value is defined manually. With the aim to solve this lack, we defined a threshold of the 40%, that will allow to stop the algorithm if the IL increases more than 40% in just one iteration.

Negative variation of *Information Lost*. The following technique introduced is the detection of IL as a negative value. This happens when the difference between the best instance and the worst instance it is outsize and the best instance it is exceeding its target. As we have done previously, we have created two levels to categorize the IL when it is negative.

The first level is assigned when IL is between −10% and −30%. In this case we notify the instance that it is exceeding its target value. The second level is the critical level and is assigned when IL is greater than −30%. In this case we will stop the algorithm and report the instance that is having an unusual behaviour.

5   Implementation

Our architecture can be shown as five components that conform our application and that are represented in Fig. 3. The shaded elements are the new components of the architecture, they are the extension performed taking as the base the framework proposed in [10].

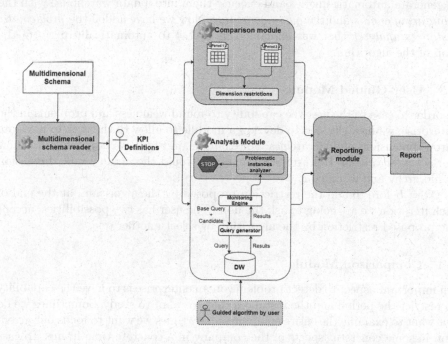

Fig. 3. Architecture of the analysis system

5.1   Multidimensional Schema Reader

The first module of our application is a multidimensional schema reader that it is able to read a multidimensional schema in order to determine the dimensions, hierarchies and levels of the schema that we have defined.

The module needs an XML file with the description of the multidimensional schema. From the XML file, the module is able to extract the different dimensions, the hierarchies of the dimensions and the levels defined in each hierarchy. Once we have identified the dimensions hierarchies and levels, the module will look for the measures defined and the type of the aggregation (sum, count, etc.)

5.2 Analysis Module

Once we have read the multidimensional schema, the user should declare the values for at least one of the measures detected in order to be able to run the algorithm looking for deviations. The values defined in a measure are:

- **Target value:** The value that the measure should achieve.
- **Threshold value:** The value between the target and the worst value.
- **Worst value:** The lowest value that a measure can have.

Analysis module was extended from [10], previously this element was able to generate automatic queries and execute them into a data warehouse with the *monitoring engine* and the *query generator*. Now we have added the *problematic instances analyzer* that was explained in Sect. 4 to automatically perform the stop of the algorithm.

5.3 User-Guided Module

In order to ease to the user the capability to found weakness and problems in the enterprise, we have decided to develop a module to allow to the user to navigate throw dimensions and hierarchies. For example, in Fig. 2, we have explained how works the algorithm. However sometimes we would choose a specific dimension or hierarchy and examine it.

Now, before to add the restriction proposed by the algorithm, at the end of each iteration we are going to ask the user. The user has two possibilities, accept the proposed restriction by the algorithm or select another one.

5.4 Comparison Module

An important aspect to detect problems on an enterprise is to have the capability to restrict the period and the instances that we want to study. Sometimes we do not want to examine the entire company, many times we want to focus our attention in some concrete aspects of the company in a concrete time frame. To ease this task, we have implemented a module that is able to perform a comparison in two different time frames and search problems in concrete instances.

In order to perform this analysis we need a time dimension defined to perform the time restriction, once we have identified the time dimension, we ask the user if he wants to restrict the analysis to some concrete instances, if the user accepts he will be able to limit the search in concrete dimensions, hierarchies and levels. As an example, in a large company, we are able to compare the sales in *Spain* in the first quarter against the sales in *France* in the second quarter.

5.5   Reporting Module

Finally, once we have performed our analysis, the framework creates a brief report with the results obtained. For example, if we perform a comparison analysis as we explained before, the report generated informs which of the time frames was better and the difference between them in percentage. We show below an example of the report generated by the execution of the framework over a sales example multidimensional model:

For the measure Quantity in period 1, there were 443 and in period 2 there were 2060. This means that the second period was better, concretely the second period was 365.01% better than the first period.

6   Evaluation

In order to test the described system we have performed a real case study of an online betting house, data provides the bets that users of the platform have done during 2014 and 2016. In Fig. 4 we can see the multidimensional model used to test the system.

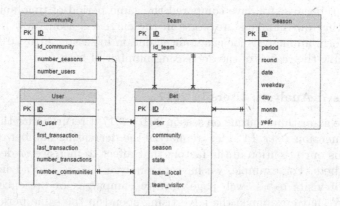

Fig. 4. Multidimensional model of the case study

The KPI that we are going to examine is the *number of bets*. So for the first test we are not going to define any restriction, just we are going to take all the data that we have available and we are going to define an IL threshold of the 80% and our the target value of bets will be 16.000.

The result obtained after the execution of the framework is the halt of the algorithm in the first iteration, notifying that the restriction performed is the dimension *Season*. In this dimension we have two hierarchies and the framework has selected the hierarchy called *Season*.

6.1   General Analysis: The Season Dimension

In this case, the module is warning us about a problem with the IL, the level of severity of IL is critical because the IL calculated in the first iteration is -2.200%.

The framework reports that there is an anomaly in data, concretely in season 2015–2016. When we examine data we detect that there is a great anomaly, season 2015–2016 has 14.490 bets against 1.215 bets from season 2016–2017 and as we have defined a target value of 16.000 bets, the target for each season (sub-KPI) is 8.000 and as we can check one of the seasons is exceeding their target and the other instance, season 2016–2017, is failing. The reason for this anomaly is due season 2016–2017 is not finished, looking in the data, for this season we have data until October 2016, so the reason for the large difference between the two seasons is due we do not have compared two complete seasons.

If we want to compare the evolution in the number of bets across seasons (2015–2016 and 2016–2017), we should use the *comparator module* that we have described previously. With this utility, if we select the periods of time where we have data in both seasons (August and September) we can perform the analysis. The output of the framework is that in the same period of time, in 2016–2017 we are obtaining a number of bets 55% lower than in 2015–2016. This report is very useful because we have compared the same period of time and we have discovered that our numbers are worse in the current season. Now if we want, we can perform an analysis focusing our attention in season 2016–2017 in order to try to know the reason of the decreasing number of bets.

6.2   Focused Analysis: Users in 2017

Running the algorithm focusing on season 2016–2017 it returns that the problem is in the dimension *User*. In order to analyze the decreasing number of bets, we have to focus our attention in the factors that affect to increase or decrease the number of bets. For example, we have to analyze factors like ad investment, a number of visits to the web page, loyalty campaigns or presence in social networks. We have examined the advertising spend in the same period and we have discovered that the spend in the current season has decreased a 70% from season 2015–2016.

Finally, we can conclude that one of the main factors that are triggering that we are not accomplishing our targets, it is because we have done a large cutout in ad investment, that in an online business with a lot of competitors has a great impact.

7   Conclusions and Future Work

We have presented an approach to detect problems and help enterprises understand where their weaknesses originate from, thereby enabling them to take corrective actions. We have explained the theoretical foundations of the framework and the rules that lead to the detection of different problems within KPIs, providing a flexible and more comprehensive analysis of the situation. We have implemented our approach and tested it in a real case, finding out weaknesses and needs for improvement that lead to a better business strategy.

As a future work, we are focused on improving the analysis process in order to both deal with larger quantities of data and find more insights. We are going to work in a prototype with the main goal of parallelizing searches in order to avoid only to look in the worst instance, this search will be more complex and accurate with the aim of improve the discovery of the problem. Additionally, we want to improve the interpretability of the results, by developing a visualization module that helps to interact with the solution and to better understand how weaknesses detected relate to business goals.

Acknowledgements. This work has been partially funded by the Spanish Ministry of Economy and Competitiveness (MINECO/FEDER) under the Granted Project SEQUOIA-UA (Management requirements and methodology for Big Data analytics) (TIN2015–63502-C3-3-R). This work has been partially funded by the University of Alicante under a R&D grant for initiation to research (BOUA 28/06/2016).

References

1. Kucukaltan, B., Irani, Z., Aktas, E.: A decision support model for identification and prioritization of key performance indicators in the logistics industry. Comput. Hum. Behav. **65**, 346–358 (2016)
2. Jeusfeld, M.A., Thoun, S.: Key performance indicators in data warehouses. In: Zimányi, E., Abelló, A. (eds.) eBISS 2015. LNBIP, vol. 253, pp. 111–129. Springer, Cham (2016). doi:10.1007/978-3-319-39243-1_5
3. Dyson, R.G.: Strategic development and swot analysis at the university of warwick. Eur. J. Oper. Res. **152**(3), 631–640 (2004)
4. Hill, T., Westbrook, R.: Swot analysis: it's time for a product recall. Long Range Plann. **30**(1), 46–52 (1997)
5. Schoemaker, P.J., van der Heijden, C.A.: Integrating scenarios into strategic planning at royal dutch/shell. Plann. Rev. **20**(3), 41–46 (1992)
6. Kerzner, H.R.: Project Management Metrics, KPIs, and Dashboards: A Guide to Measuring and Monitoring Project Performance. Wiley, Hoboken (2011)
7. Maté, A., Trujillo, J., Mylopoulos, J.: Stress testing strategic goals with SWOT analysis. In: Johannesson, P., Lee, M.L., Liddle, S.W., Opdahl, A.L., López, Ó.P. (eds.) ER 2015. LNCS, vol. 9381, pp. 65–78. Springer, Cham (2015). doi:10.1007/978-3-319-25264-3_5
8. Sarawagi, S.: User-adaptive exploration of multidimensional data. VLDB **2000**, 307–316 (2000)
9. Kaplan, R.S., Norton, D.P.: Putting the balanced scorecard to work. Perform. Measur. Manage. Apprais. Sourceb. **66**, 17511 (1995)
10. Maté, A., Zoumpatianos, K., Palpanas, T., Trujillo, J., Mylopoulos, J., Koci, E.: A systematic approach for dynamic targeted monitoring of kpis. In: Proceedings of 24th Annual International Conference on Computer Science and Software Engineering, IBM Corp. pp. 192–206 (2014)

OntoCom 2017 - 5th International Workshop on Ontologies and Conceptual Modeling

Preface

This volume collect articles presented at the fifth edition of the International Workshop on Ontologies and Conceptual Modeling (OntoCom 2017). It was held in the context of the the 36th International Conference on Conceptual Modeling in Valencia, Spain. OntoCom 2017 concerns the practical and formal application of ontologies to conceptual modelling. We invite leading ontologists and conceptual modelers to discuss and comparatively analyze different foundational ontologies, their meta-ontological choices and their associated modelling processes. The workshop consists out of two different aspects: first, the workshop focuses facilitating discussion to discover and evaluate how different ontologies would distinguish themselves in representing different kinds of case studies that each represent a modelling problem. Second, the workshop hosts a paper session where researchers can present their current research efforts and/or will be able to discuss with experienced academics new research ideas.

For this edition we received 6 submissions from Belgium, Brazil, Colombia, Germany, Hungary, United Kingdom and Spain. These papers were reviewed by at least two members of the program committee and eventually two papers were accepted for publication.

In the paper *"Automatically Annotating Business Process Models with Ontology Concepts at design-Time"*, Dennis M. Riehle, Sven Jannaber, Patrick Delfmann, Oliver Thomas and Jörg Becker propose an automatic approach for annotating process models using a thesaurus, naming conventions and a domain ontology.

In *"OPL-ML: A Modeling Language for Representing Ontology Pattern Languages"*, Glaice Quirino, Monalessa Barcellos and Ricardo A. Falbo, present OPL-ML, a visual modeling language for representing OPLs. It aims to overcome the inconsistencies in current representations, which have been identified in the literature. It was designed according to the principles of the Physics of Notation (PoN), and following the design process defined by PoN Systematized (PoN-S).

We would like to thank the authors who considered Onto.com and the program committee members for their reviews. Finally, we would like to thank the ER 2017 workshop chairs and organization committee for giving us the opportunity to organize the workshop.

August 2017

Frederik Gailly
Giancarlo Guizzardi
Mark Lycett
Chris Partridge
Michael Verdonck

Automatically Annotating Business Process Models with Ontology Concepts at Design-Time

Dennis M. Riehle[1(✉)], Sven Jannaber[2], Patrick Delfmann[3], Oliver Thomas[2], and Jörg Becker[1]

[1] European Research Center for Information Systems, University of Münster, Münster, Germany
{dennis.riehle,joerg.becker}@ercis.uni-muenster.de
[2] Institute for Information Management and Information Systems, University of Osnabrück, Osnabrück, Germany
{sven.jannaber,oliver.thomas}@uni-osnabrueck.de
[3] Institute for Information Systems Research, University of Koblenz-Landau, Koblenz, Germany
delfmann@uni-koblenz.de

Abstract. In business process modelling, it is known that using a consistent labelling style and vocabulary improves process model quality. In this regard, several existing approaches aim at the linguistic support for labelling model elements. At the same time, domain-specific ontologies have been proposed and used to capture important process-related knowledge. However, these two areas are largely disconnected up to now. Although some research suggests annotating ontology concepts to process models, for instance, to interpret and reason about a process model, annotation has not yet gained traction in practice since it still has to be done in a highly manual effort. We thus provide an automated, language-independent methodology for using labelling assistance functionalities to identify and annotate relevant ontology concepts to process model elements using a four-step procedural model.

Keywords: Process models · Ontology · Automatic annotation · Analysis

1 Motivation

In information systems, conceptual models are frequently used within research and practice. They enable a graphical and structural representation of the main concepts of an application domain and their relations [1]. One type of conceptual models is business process models, which represent the set of activities performed within a business process together with their logical execution order and environment. Business process models support the analysis of business processes, which is a common task of business process management [2]. Over the last decades, business process management (BPM) has been of growing interest. Researchers have developed new approaches and algorithms to further improve BPM techniques and practitioners have implemented BPM techniques in their organisations [3]. Not surprisingly, organisations nowadays maintain large repositories of process models, consisting of hundreds if not thousands of models. A three-study at Suncorp-Metway Ltd (an Australian insurer) dealt with over 6000 process

© Springer International Publishing AG 2017
S. de Cesare and U. Frank (Eds.): ER 2017 Workshops, LNCS 10651, pp. 177–186, 2017.
https://doi.org/10.1007/978-3-319-70625-2_17

models, after the organisation had gone through several mergers and acquisitions [4]. Such process models are usually created by modellers, that is, human beings. Therefore, the creation of process models is influenced by the individual perception of the modellers and, hence, different modellers may create models of different quality. Modelling expertise of modellers has been shown to influence the comprehensibility of models [5]. However, different models may still name the same things differently, leading to naming conflicts [6]. The more people there are involved in the modelling process, the more it is likely that resulting process models differ largely in respect to the used terminology [7]. Still modellers appreciate a consistent naming of model elements [8].

While there are different approaches to prevent naming conflicts (see section 'Related Work'), these approaches usually only improve the labelling of process models (e.g., by defining a syntax for natural language), but two different modellers do still not necessarily refer to the same semantic concept. Consequently, the domain specific concepts need to be shared among all modellers, using an ontology to represent the body of domain knowledge [9]. An ontology is an engineering artefact, which includes not only a vocabulary to describe reality, but also includes a set of intended meanings for concepts and vocabulary included in the ontology [10].

To connect process models with domain ontologies, elements from the process models need a relation to concepts from the ontology. Given that the ontology contains organisational knowledge, this allows an advanced business process analysis. For example, organisational compliance regulations regarding process execution could be stored in the ontology and then, during analysis, be used to check business process models for compliance. Currently, the relation between an ontology concept and a process element needs to be established manually, which is a challenging and resource consuming task given the size of current process model repositories. Therefore, this paper presents a methodology to automatically create the relations between elements in a business process model and concepts from an ontology during design-time. Reducing the administrative overhead to link process models and ontologies may foster the use ontology-based process analysis in the future.

This paper is structured as follows: First, we shortly discuss related work. Then, we discuss the research gap that our methodology aims to close. After this, we introduce our methodology automatic annotations. Finally, we consider limitations of our work and provide a conclusion and outlook.

2   Related Work

Early research towards terminological standardisation proposes glossaries and structural rules to achieve unified naming. A rule for activities in business process might be, for example, "<Verb, Imperative> <Noun>", which would make "process order" a valid label, but not "order is processed". More advanced approaches like [11] suggest to use a domain ontology instead of the glossary. A partially automatic methodology for copying concept labels from ontologies into models modelled with Business Model and Notation (BPMN) has been introduced by [12]. However, their approach requires states that business objects can undergo and has been applied only to BPMN so far. Another approach to match elements from conceptual models with concept from formal semantic schemata is provided by [13], but their approach does not provide the modeller with suggestions at design-time. Several studies (e.g., [14, 15]) use online dictionaries likes WordNet to suggest labels for model elements.

Regarding terminological standardization, we can distinguish between approaches which provide terminological standardisation at design time and those, which check for terminological standardisation as part of a process analysis procedure. For process models which are not terminologically standardized, there are several approaches in the literature, which judge the quality of process element labels and give hints on possible naming violations (e.g., [16]) or resolve them (e.g., [17]). Compared thereto, approaches which support process modellers at design time are rather young. A technique which provides suggestions to the modeller has been suggested by [18]. Their approach uses a dictionary with a vocabulary that contains all necessary words to label process elements, the so-called domain thesaurus. By using structural rules they provide the modeller with terminologically standardised labels at design time. This approach has the advantage of supporting the creation of correct process models right from scratch, while still being applicable to existing process models as well.

A recent study presents an automatic annotation of process models with concepts from a taxonomy [19]. While this approach tracks a similar target, it is different in the manner that it computes distributional similarities between process model elements and taxonomy concepts to detect inconsistent terminology. This can lead to wrong results, especially if labels are not terminologically standardised beforehand. Additionally, this approach does not provide an automatic annotation at design-time.

Summing up, there are several techniques to achieve terminological standardisation, some at design time, some as part of process analysis. However, up to now, no approach is capable of automatically annotating ontology concepts to process models at design time. Therefore, annotation of ontologies requires a prohibitively high manual effort and annotation is rarely used in practice. Consequently, existing approaches utilizing domain ontologies during the process analysis and improvement cannot gain their full potential. To close this research gap, we thus propose a methodology to automatically annotate conceptual models with ontology concepts at design-time.

3   Automatic Annotation of (Process) Models with Ontologies

Our methodology aims to achieve essentially two things: First, terminological standardisation shall be assured, for which we follow the methodology provided by [18]. Second, elements from process models shall be automatically related to concepts from an ontology to foster a common understanding of process elements and to enable further ontology-based process analyses and process modelling assistance in the future.

3.1   Terminological Standardization

A key requirement for automatic annotation are standardised and unambiguous identifiers for all elements within a business process model. This is achieved by ensuring terminological standardisation of all elements' labels at design-time. For this, essentially two things are required (cf. Fig. 1):

1. A domain thesaurus with valid words (ideally corresponding to the labels of the ontology's concepts to easily establish the annotation)
2. Syntactical naming conventions

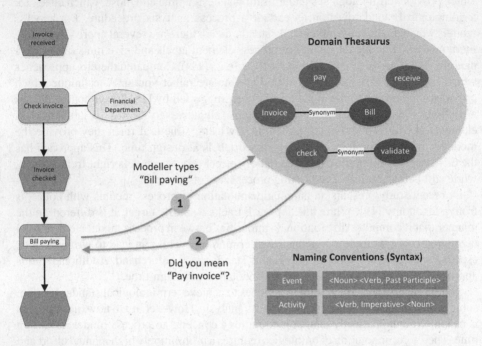

Fig. 1. Terminological standardization at design-time [18] (Color figure online)

The domain thesaurus contains a vocabulary of all words of a natural language considered valid in the respective application domain. Words in the domain thesaurus are declared as being a noun or a verb and need to be in their uninflected form, which is the singular form for nouns and the infinitive for verbs. Furthermore, the domain thesaurus includes relations between words, for example synonyms, homonyms or antonyms. An exemplary thesaurus could include the information of "bill" being a synonym for "invoice", "invoicing" being a word formation of "invoice" and "correct" being an antonym of "incorrect".

While lexical databases like WordNet can easily be adapted rather than building a domain thesaurus from scratch, the domain thesaurus needs to contain additional, domain-specific information. In case of synonyms, one of the synonyms has to be marked as dominant, in order to specify which of the synonyms is to be used preferably within the application domain. Dominant words are shown with a blue background in the domain thesaurus of Fig. 1, while non-dominant words are shown with a grey background.

Next, syntactical naming conventions are required. These naming conventions follow the suggestions by [20]. Such naming conventions differ depending on the type of element which is to be named. In the area of process modelling, typical element types are activities, events or organisational units. For example, activities could be named by the rule "<Verb, Imperative> <Noun>", which would make "Write paper" a syntactically correct label, while "Writing paper" or "Paper is written" would be syntactically incorrect labels. Note that there can be more than one syntactical naming convention per element type. For instance, activities in process models may require more complex phrases than the one mentioned above.

Using the domain thesaurus and the syntactical naming conventions, an automatic suggestion of terminologically standardised labels can be supported within any modelling tool. In concordance with [18], this works as follows: First, the phrase entered by the modeller (cf. Step 1 in Fig. 1) needs to be parsed into uninflected words. Single words have to be recognised and turned into their corresponding lexeme, which is the singular form for nouns or the infinitive for verbs. Software for this is already known in the literature (see [18] for an overview). These lexemes are then looked-up in the domain thesaurus in order to resolve synonyms to their dominant terms. In case a dominant term cannot be found because, for example, the term entered by the modeller is not known to the thesaurus, a further automatic search can be performed in general lexicons like wordnet, which may return synsets that contain a term known to the thesaurus. In the example shown in Fig. 1, the modeller entered "Bill paying", which would result in the lexemes "bill" and "pay", of which the premier would resolve into the dominant term "invoice" after consulting the domain thesaurus.

With the standardised words and the syntactical naming conventions, the phrase can be reformulated based on the element type the original phrase was typed into. Again, in the example shown in Fig. 1, the modeller typed a label for an activity, therefore the rule "<Verb, Imperative> <Noun>" is to be used. With the words "invoice" and "pay" this results in the phrase "pay invoice". This phrase is suggested to the modeller as a correct and valid labelling (cf. Step 2 in Fig. 1).

3.2 Domain Ontology

The second key requirement for automatic annotation of process models is the presence of a domain ontology, which represents all concepts of the organisation. Such an ontology also includes elements and mutual relations (so far similar to the domain thesaurus). In addition, ontology elements are semantic concepts rather than simple words, and relations can be semantically typed (e.g., "requires", "is a", "targets at"). These semantic relationships are used to express interrelations between concepts that do not necessarily occur in process models, organisational charts, data models or the like. Therefore, they can be used to depict advanced domain semantics. Furthermore, by using relations defining inheritance, it is possible to store abstract instances (often called classes, stored in the so-called TBox of the ontology) and concrete instances (stored in the so-called ABox of the ontology) in one and the same ontology [21].

The following example consists of a small organisation, which has two departments and two processes (see Fig. 2). Two abstract classes are defined: "departments" and "processes", which is achieved with the two prefixes defined in the second and third line of the listing below. Furthermore, there are two instances of the process class: "pro:invo_check", which represents the task of checking an invoice and "pro:invo_pay", which represents the task of paying an invoice. In Addition, there are two instances of the department class: "org:dep_fin" is the financial department and "org:dep_hr" is the human resources department. The financial department ("org:dep_fin") is responsible for both tasks, which is defined by the relation "pro:responsible". Further relationships can be added as needed.

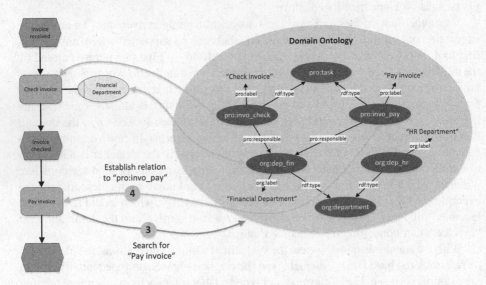

Fig. 2. Automatic annotation of ontology concepts

For all concrete concepts in the ontology, we define a label, under which the concept can be displayed to a user. One very important requirement here is that these labels use the same domain thesaurus and syntactical naming conventions as described above to allow proper annotation of process model elements. To ensure that this requirement is met, one could, for example, derive the domain thesaurus automatically from an existing ontology on the one hand, or terminologically standardise an ontology based on an existing domain thesaurus using the above-mentioned methodology.

3.3   Methodology to Realise Automatic Annotations

Relying on the two requirements discussed above, we define the following steps as our methodology to enable automatic annotations of process models at design-time:

1. A phrase entered by the modeller is parsed into lexemes, which are then looked-up in the domain thesaurus.
2. From the domain thesaurus, a terminologically standardised phrase is generated.
3. The standardised phrase is used for naming process elements and to search for related concepts in the domain ontology.

4. Matching concepts are proposed to the modeller, who can decide whether or not the proposed concepts are appropriate for annotation. If so, the annotation link is established automatically.

Step 1 and 2 are described in Sect. 3.1 (see also Fig. 1). These two steps ensure terminological standardisation and are in harmony with the methodologies already described in literature. Step 3 and 4 realise the automatic annotation itself.

If the modeller accepts one of the suggested phrases, this phrase is used to query the domain ontology for a matching concept (cf. Step 3 in Fig. 2). Since labels in the ontology are already terminologically standardised, searching for related concepts is left to a simple string comparison. Considering the sample process shown in Sect. 3.1 and the sample ontology presented in Sect. 3.2, the follow-up steps 3 and 4 are shown in Fig. 2. With the example phrase "pay invoice", the ontology concept "pro:invo_pay" is found, which describes the task of paying an invoice and hence matches the activity the modeller wants to label. In the last step, a link between the matching ontology concept "pro:invo_pay" and the process element is created automatically if the modeller accepts the suggested annotation (cf. Step 4 in Fig. 2). Finally, the function "Check invoice" is linked to the concept "pro:invo_check", the organisational unit "Financial department" is linked to the concept "org:dep_fin" and the function "Pay invoice" is linked to the concept "pro:invo_pay" from the domain ontology.

While searching for concepts with the exact same title is rather straight-forward, more sophisticated searches can be applied as well. Due to the use of terminological standardisation, we can also search for individual lexemes within the ontology. This allows to also allows to suggest the user with concepts that do not fit their process element exactly, but to provide them with additional concepts known in the organisation that could fit the process. In consequence, it is also possible to annotate multiple concepts of different types, establishing relations to all process-related knowledge present in the organisation.

4 Limitations and Outlook

While our methodology provides automatic annotation of business processes with concepts from an ontology, it is limited to the presence of a terminologically standardised ontology. However, since ontologies are usually modelled as well (though probably not by process modellers), the same technique (without the automatic linking) can be applied to the process of modelling the ontology, ensuring a terminologically standardised ontology during creation already. Similarly, the technique could also be applied to already existing ontologies, which, however, involves human interaction to fix invalid, non-terminologically-standardized labels in the ontology.

A further limitation, which needs to be regarded more critically, is the fact that ontologies actually need to be "useful" to foster a common understanding of domain concepts. Such usefulness highly depends on the actual ontology and its contents. It is easy to state that a useful ontology should contain all necessary information, while not specifying which information are actually necessary. For an organisation, this might be all tasks performed within the organisation, all actors and stakeholders, all inter-process dependencies and all responsibilities. It is important that the information present in the ontology is correct, accurate and most importantly complete.   •

Until now, we have not yet discussed an ontology-based analysis of process models. Through the introduction of our automated annotation, links between process models and ontologies no longer need to be created manually by means of extensive human work. The simplicity with which such links can now be created enables a whole new spectrum of business process analyses. Let us start with an example. Besides what is contained in the sample ontology presented in Sect. 3.2, tasks, departments and responsibilities, a useful ontology might include the information that an invoice needs to be checked before in can be paid. This literally means that the business process "Check invoice" is required to be executed before the business process "Pay invoice" can be executed. Other useful information an ontology could include, are goals which a process reaches when executed. This has already shortly been mentioned previously, when the process of submitting an invoice at the financial department leads to reaching the goal of submitting an invoice.

Keeping such ontology in mind, we can regard new aspects in the process of process modelling as well as in the process of analysing process models. Sticking to the example that invoices should only be paid if they have been checked previously, a process modeller could be given constructional assistance, i.e. by automatically suggesting them the task "pay invoice" after they have placed the task "check invoice" in their process model. While there are already a lot of papers on so-called recommender systems (e.g. [22–25]), these approaches currently learn from existing process models. Consequently, the quality of such suggestions falls with the quality of the existing models. Introducing a domain ontology to generation modelling suggestions from, bears great potential for better modelling suggestions.

Furthermore, process models could also be checked for compliance with business regulations or for inefficiencies or general flaws automatically. For this, compliant task sequences could be added to the ontology and query languages could be used to validate compliance. Besides inter-process dependencies, responsibilities could be validated as well. In the example shown in Fig. 1, a suitable algorithm could notify the modeller that the activity "Pay invoice" ("Bill paying" before terminological standardization) is missing the organizational unit of the financial department, since the ontology has the information that the financial department is necessary for the task of paying an invoice.

While both the recommender system as well as the compliance analysis require a well-modelled ontology, we argue that due to the reusability of ontology in many different aspects, companies will be more willing to spend time and effort in creating a domain ontology of high quality. With our prototypical implementation, we have shown that our methodology actually works and can be used to work with at the design-time of process modelling. We are aware that this is only a limited evaluation, as we have applied it to artificial data only. In future research, we plan to apply our methodology to real-world processes and ontologies to see if it also works in practice. Since an in-depth evaluation requires more space, we plan to publish them as a separate paper including more details about experiences applying our methodology in practice.

Lastly, further research might also further investigate the domain thesaurus. Currently, we consider the domain thesaurus as an organization-dependant artefact, but it might be possible the reuse the domain thesaurus across different organisations. Researchers should analyse to which extent the thesaurus can be reused, across different organisations in the same industry or probably also across organisations in different industries. A high reusability would be beneficial under economic aspects.

5 Conclusion

We have presented a methodology to automatically annotate business process models with ontology concepts at design-time, by creating links between the process models and the ontology through terminological standardisation. Significantly reducing the effort which currently needs to be put into this by establishing these links manually, our methodology has the potential to improve future process modelling. Relations between ontology concepts and process models do not only help modellers of multi-national organisations to share a common understanding or business process models, but also enable further modelling support techniques and process model analyses, which – in this form – have not been possible with the techniques known in literature before.

While our paper has a conceptual perspective on automatic annotation, we have already implemented a prototypical artefact and used our methodology on artificially created models. In addition, we have shown areas for further research, namely to examine linguistic modelling assistance from a technical perspective, with the goal of providing and evaluating different algorithms for a possible implementation in the future. With our outlook to ontology-based model analysis, we have shown that our work sets the base for complex and advanced analyses in the area of compliance checking, which is an area that becomes more and more important for many organisations.

References

1. Schreiber, G., Akkermans, H., Anjewierden, A., Hoog, R.D., Shadbolt, N.R., Wielinga, B.: Knowledge Engineering and Management: The CommonKADS Methodology. MIT Press, Cambridge (2000)
2. Weske, M.: Business Process Management: Concepts, Methods, Technology. Springer, Berlin (2007)
3. van der Aalst, W.M.P.: Business process management: a comprehensive survey. ISRN Softw. Eng. **2013**, 1–37 (2013)
4. La Rosa, M., Dumas, M., Uba, R., Dijkman, R.M.: Business process model merging: an approach to business process consolidation. ACM Trans. Softw. Eng. Methodol. TOSEM, **2**(2), 11:1–11:42 (2012)
5. Mendling, J., Strembeck, M., Recker, J.: Factors of process model comprehension-findings from a series of experiments. Decis. Support Syst. **53**, 195–206 (2012)
6. Batini, C., Lenzerini, M.: A methodology for data schema integration in the entity relationship model. IEEE Trans. Softw. Eng. **SE-10**, 650–664 (1984)
7. Hadar, I., Soffer, P.: Variations in conceptual modeling: classification and ontological analysis. J. Assoc. Inf. Syst. **7**, 20 (2006)
8. Fellmann, M., Zarvic, N., Metzger, D., Koschmider, A.: Requirements catalog for business process modeling recommender systems. In: 12th International Conference on Wirtschaftsinformatik (WI 2015), pp. 393–407 (2015)

9. Studer, R., Benjamins, V.R., Fensel, D.: Knowledge engineering: principles and methods. Data Knowl. Eng. **25**, 161–197 (1998)
10. Guarino, N.: Formal ontology and information systems. In: Proceedings of the FOIS 1998, Trento, Italy, pp. 3–15 (1998)
11. Greco, G., Guzzo, A., Pontieri, L., Saccà, D.: An ontology-driven process modeling framework. In: Galindo, F., Takizawa, M., Traunmüller, R. (eds.) DEXA 2004. LNCS, vol. 3180, pp. 13–23. Springer, Heidelberg (2004). https://doi.org/10.1007/978-3-540-30075-5_2
12. Born, M., Dörr, F., Weber, I.: User-friendly semantic annotation in business process modeling. In: Weske, M., Hacid, M.-S., Godart, C. (eds.) WISE 2007. LNCS, vol. 4832, pp. 260–271. Springer, Heidelberg (2007). https://doi.org/10.1007/978-3-540-77010-7_25
13. Fill, H.-G., Schremser, D., Karagiannis, D.: A generic approach for the semantic annotation of conceptual models using a service-oriented architecture. Int. J. Knowl. Manag. **9**, 76–88 (2013)
14. Rizopoulos, N., Mçbrien, P.: A general approach to the generation of conceptual model transformations. In: Pastor, O., Falcão e Cunha, J. (eds.) CAiSE 2005. LNCS, vol. 3520, pp. 326–341. Springer, Heidelberg (2005). https://doi.org/10.1007/11431855_23
15. Bögl, A., Kobler, M., Schrefl, M.: Knowledge acquisition from EPC models for extraction of process patterns in engineering domains. In: Proceedings of the 4th Multikonferenz Wirtschaftsinformatik (MKWI 2008), pp. 1601–1612 (2008)
16. Leopold, H., Eid-Sabbagh, R.H., Mendling, J., Azevedo, L.G., Baião, F.A.: Detection of naming convention violations in process models for different languages. Decis. Support Syst. **56**, 310–325 (2013)
17. Pittke, F., Leopold, H., Mendling, J.: Automatic detection and resolution of lexical ambiguity in process models. IEEE Trans. Softw. Eng. **41**, 526–544 (2015)
18. Delfmann, P., Herwig, S., Lis, L.: Unified enterprise knowledge representation with conceptual models - capturing corporate language in naming conventions. In: Proceedings of the 30th International Conference on Information Systems (ICIS 2009) (2009)
19. Leopold, H., Meilicke, C., Fellmann, M., Pittke, F., Stuckenschmidt, H., Mendling, J.: Towards the automated annotation of process models. In: Zdravkovic, J., Kirikova, M., Johannesson, P. (eds.) CAiSE 2015. LNCS, vol. 9097, pp. 401–416. Springer, Cham (2015). https://doi.org/10.1007/978-3-319-19069-3_25
20. Kugeler, M., Rosemann, M.: Fachbegriffsmodellierung für betriebliche Informationssysteme und zur Unterstützung der Unternehmenskommunikation. Informationssystem Archit. Fachaussch. **5**, 8–15 (1998)
21. Giacomo, G.D., Lenzerini, M.: TBox and ABox reasoning in expressive description logics. In: Proceedings of the 5th International Conference on Principles of Knowledge Representation and Reasoning (KR 96) (1996)
22. Chan, N.N., Gaaloul, W., Tata, S.: Assisting business process design by activity neighborhood context matching. In: Liu, C., Ludwig, H., Toumani, F., Yu, Q. (eds.) ICSOC 2012. LNCS, vol. 7636, pp. 541–549. Springer, Heidelberg (2012). https://doi.org/10.1007/978-3-642-34321-6_38
23. Koschmider, A., Hornung, T., Oberweis, A.: Recommendation-based editor for business process modeling. Data Knowl. Eng. **70**, 483–503 (2011)
24. Li, Y., Cao, B., Xu, L., Yin, J., Deng, S., Yin, Y., Wu, Z.: An efficient recommendation method for improving business process modeling. IEEE Trans. Ind. Inform. **10**, 502–513 (2014)
25. Smirnov, S., Weidlich, M., Mendling, J., Weske, M.: Action patterns in business process model repositories. Comput. Ind. **63**, 98–111 (2012)

OPL-ML: A Modeling Language for Representing Ontology Pattern Languages

Glaice K.S. Quirino[1,2], Monalessa P. Barcellos[1(✉)], and Ricardo A. Falbo[1]

[1] Ontology and Conceptual Modeling Research Group (NEMO),
Department of Computer Science, Federal University of Espírito Santo,
Vitória, ES, Brazil
glaice.monfardini@ifes.edu.br,
{monalessa, falbo}@inf.ufes.br
[2] Federal Institute of Espírito Santo, Cachoeiro de Itapemirim, ES, Brazil

Abstract. Reuse has been pointed out as a promising approach for Ontology Engineering. Reuse in ontologies allows speeding up the development process and improves the quality of the resulting ontologies. The use of patterns as an approach to encourage reuse has been explored in Ontology Engineering. An Ontology Pattern (OP) captures a solution for a recurring modeling problem. Very closely related OPs can be arranged in an Ontology Pattern Language (OPL). An OPL establishes relationships between the patterns and provides a process guiding the selection and use of them for systematic problem solving. To make it easier using an OPL, the relationships between the patterns and the process for navigating them should be represented in a clear and unambiguous way. A visual notation can be used to provide a visual representation of an OPL, aiming at improving communication. To facilitate understanding an OPL and strengthen its use, this visual notation must be cognitively rich. This paper presents OPL-ML, a visual modeling language for representing OPLs.

Keywords: Ontology Pattern Language · Ontology · Visual notation · Visual modeling language

1 Introduction

Nowadays, ontology engineers are supported by a wide range of methods and tools. However, building ontologies is still a difficult task. In this context, an approach that has gained increasing attention in recent years is the systematic application of *ontology patterns* (OPs), which favors the reuse of encoded experiences and promotes the application of quality solutions already applied to solve similar modeling problems [1–4]. An OP describes a recurring modeling problem that arises in specific ontology development contexts, and presents a well-proven solution for this problem [2]. Experiments, such as the ones presented in [5], show that ontology engineers perceive OPs as useful, and that the quality and usability of the resulting ontologies are improved.

S. de Cesare and U. Frank (Eds.): ER 2017 Workshops, LNCS 10651, pp. 187–201, 2017.
https://doi.org/10.1007/978-3-319-70625-2_18

As pointed by Blomqvist et al. [3], although the ideas behind OPs are not completely realized in practice yet, the process is ongoing. As OPs become more mature and the community collects more and more experience using them, a situation that already occurs in Software Engineering, where patterns have been studied and applied for a long time, will emerge in Ontology Engineering. In a pattern-based approach to Ontology Engineering, several patterns can be combined to derive a new ontology. Such approach requires the existence of a set of suitable patterns that can be reused in the development of new ontologies, and a proper methodological support for selecting and applying these patterns [4]. In this context, we need to record how different patterns relate to each other in a more abstract level, and thus we need an ontology pattern representation language [3]. The first steps towards developing such a representation language have already been undertaken, giving rise to Ontology Pattern Languages (OPLs) [6, 7]. An OPL is a network of interconnected OPs that provides holistic support for solving ontology development problems. An OPL contains a set of inter-related OPs, plus a modeling process guiding on how to use and combine them in a specific order [6].

To facilitate ontology engineers understanding an OPL, the relationships between the patterns, as well as the process for navigating them, should be represented in a clear way. Ideally, such representation should be visual, since visual representations are effective, by tapping into the capabilities of the human visual system [8].

The use of OPLs is a recent initiative. There are still only few works defining OPLs, all of them from the same research group: the Ontology and Conceptual Modeling Research Group (NEMO). This research group has developed the following OPLs [7]: Software Process OPL (SP-OPL), ISO-based Software Process OPL (ISP-OPL), Enterprise OPL (E-OPL), Measurement OPL (M-OPL), and Service OPL (S-OPL). Since these works were done by the same research group, they share commonalities. All of them use extensions of UML activity diagrams for representing the OPL modeling processes [7]. However, there are still inconsistencies and problems in such representations, as we could perceive by applying two experimental studies. Trying to overcome these problems, we developed OPL-ML, an OPL Modeling Language. For developing OPL-ML, we rely on the results of a systematic mapping of the literature that investigated visual notations for Software Pattern Languages. Moreover, OPL-ML was designed according to the principles of the Physics of Notation (PoN) [8], and following the design process defined by PoN-S (PoN Systematized) [9].

This paper aims at presenting OPL-ML, and is organized as follows: Sect. 2 discusses patterns, OPs, pattern languages and OPLs; Sect. 3 discusses the Design Science [10] methodological approach we followed to develop OPL-ML; Sect. 4 presents OPL-ML; Sect. 5 discusses a preliminary evaluation of OPL-ML; Sect. 6 concerns related works, and Sect. 7 presents our final considerations.

2 From (Ontology) Patterns to (Ontology) Pattern Languages

Patterns are vehicles for encapsulating design knowledge, and have proven to be beneficial in several areas. "Design knowledge" here is employed in a general sense, meaning design in different areas, such as Architecture and Software Engineering (SE).

In SE, for instance, patterns have been studied and applied for a long time, and there are several types of patterns, such as analysis patterns, design patterns and idioms. The main principle behind patterns is not having to reinvent the wheel.

Ontology Patterns (OPs) follow the same idea: capturing a well-proven solution for a recurring modeling problem that arises in ontology development contexts. There are also several types of OPs, such as [2]: content patterns (foundational and domain-related patterns), design patterns (logical and reasoning patterns), and idioms.

Patterns, in general, can exist only to the extent that they are supported by other patterns. There is a need to describe the context of larger problems that can be solved by combining patterns, and to address issues that arise when patterns are used in combination. This context can be provided by what in Software Engineering has been termed a *Pattern Language* (PL) [11]. According to Schmidt et al. [11], the trend in the SE patterns community is towards defining pattern languages, rather than stand-alone patterns. We have advocated that this approach should also be followed in Ontology Engineering, by defining Ontology Pattern Languages (OPLs). An OPL aims to put together a set of very closely related ontology patterns (OPs), in a system of patterns that provides, besides the OPs themselves, a process describing how to navigate, select and apply them in a consistent way. The term "pattern language" was borrowed from Software Engineering (SE). However, it is important to say that we are not talking about a language properly speaking. In "pattern language", the term "language" is, in fact, a misnomer, given that a pattern language does not typically define per se a grammar with an explicit associated mapping to a semantic domain [6].

The use of OPLs is a recent initiative. At the best of our knowledge, all existing OPLs were developed by the same research group and use extensions of UML activity diagram for representing the OPL processes [7]. However, there are still inconsistencies and problems in such representations. Thus, to help developing new OPLs, it is essential to solve these problems and to provide a well-defined modeling language for representing OPLs. The use of PLs in SE is not new. Thus, aiming to get knowledge about visual notations for PLs, we carried out a systematic mapping to investigate visual notations used to represent software PLs. A systematic mapping provides an overview of a research area, and helps identify gaps that can be addressed in future research [12].

In our systematic mapping, we searched seven digital libraries, namely: Scopus (www.scopus.com), Engineering Village (www.engineeringvillage.com), ACM (dl.acm.org), IEEE Xplore (*ieeexplore.ieee.org*), Springer (*link.springer.com*), Science-Direct (www.sciencedirect.com) and Web of Science (www.webofknowledge.com). We identified 54 software-related PLs represented by visual notations and we investigated them. Next, we summarize some results and perceptions gotten from the study.

Different elements are addressed in PLs. We identified 13 different elements, namely: pattern, pattern group, pattern subgroup, flow, mandatory flow, alternative flow, parallel outputs, parallel inputs, structural relation, optional relation, and variant patterns. Many different representations for the same element were found. For instance, we found seven different symbols for representing patterns and nine for pattern groups. Several PLs include only few elements (e.g., pattern and flow). We noticed that in most cases, PLs use less elements than necessary. Consequently, elements meaning is overloaded and their symbols tend to be cognitively poor. As a result, due to lack of clarity, it is often difficult to understand the PL.

Patterns can relate to others in different ways. The most common relations in the investigated PLs are dependency, usage and specialization. Most of the investigated PLs include dependency relations. Dependency can be understood as a broad relation and, if more specific relations are not defined, it can be not clear what dependency really means. For instance, if a pattern A depends on a pattern B, it is not clear if A depends on B because the solution given by A requires the solution given by B (thus, B must be applied before A), or because B is part of the solution given by A. In both cases, A depends on B, however, in the second case the dependency is, more specifically, a composition relation. The investigated PLs often do not explicitly define the different kinds of relations, making it difficult to understand relations among patterns and, consequently, to properly select and apply them.

Defining groups is particularly important when defining large and complex PLs. Groups can be represented in a transparent way (i.e., the patterns in a group are visible) or as black boxes (i.e., a symbol represents a pattern group and it is not possible to see the patterns inside it). The use of black boxes allows representing a PL at different detail levels, contributing to its understanding. 54% of the investigated PLs use groups to organize patterns. Only one of them use black boxes to organize the PL.

The investigated PLs use different types of models to represent their elements, providing different views to ease understanding and using the PL. Structural models present elements (e.g., pattern, patterns group) and their structural relations (e.g., dependency, composition). Process (or behavioral) models, in turn, present the possible paths to be followed to apply the patterns. 92% of the investigated studies present only one of these models (46% structural model and 46% process model). Only 8% of the PLs consider both views, half of them using a single (hybrid) model to address structural and process aspects, making it difficult for users to differentiate them.

Considering the panorama provided by the mapping study results, some gaps in the visual notations used to represent PLs can be pointed out: (i) lack of a standard visual notation; (ii) use of cognitively poor notations; (iii) lack of mechanisms to support patterns selection; and (iv) lack of a complete view of the PL, addressing both structural and process aspects.

3   Methodological Approach

For developing this work, we followed a Design Science methodological approach [10].

The addressed problem is: *How should we represent OPLs such that they become effective in guiding the use of related OPs to develop new ontologies?* The artifact to be designed is a *modeling language for representing OPLs*. A first representation already existed, the one used to represent the first versions of SP-OPL, E-OPL, ISP-OPL and M-OPL. This representation was an informal extension of the UML activity diagram (informal in the sense that there was not even a meta-model of it). Moreover, its application in each OPL was slightly different one from another. Thus, this work started by evaluating this representation to see whether it is effective in guiding the use of related OPs to develop new ontologies. A first experiment was accomplished using

ISP-OPL [13] aiming at evaluating how the guidance provided by ISP-OPL affects the productivity of ontology engineers when developing domain ontologies for SE sub-domains, and the quality of the resulting ontologies. Besides, research questions regarding the visual notation used to represent ISP-OPL were also posed. The experiment took place during the second semester of 2014, as part of the course "Ontologies for Software Engineering", an advanced course for graduate students in the Graduate Program in Informatics at Federal University of Espírito Santo, in Brazil. 19 graduate students with at least basic knowledge in conceptual modeling participated in the study. Concerning aspects related to ISP-OPL visual notation, most participants considered that it was easy to understand the OPL, but they pointed out problems related to some of the used symbols, and even more problematic, they had difficulties in following the paths in the OPL process, especially difficulties related to mandatory/optional paths.

Based on the results of this first experiment, we worked on changing some of the symbols in the visual notation, and on establishing a clearer way to use them. Those changes in the visual notation were applied in some of the OPLs existing up to that moment (E-OPL, ISP-OPL and M-OPL). The versions of such OPLs presented in [7] use this updated notation. Moreover, the updated visual notation was used to engineer S-OPL [14], and to evaluate it, we accomplished another experiment. In this study, S-OPL was used by 9 students to develop service ontologies for specific domains. The study took place as part of the course "Ontology Engineering", an advanced course for graduate students in the Graduate Program in Informatics at Federal University of Espírito Santo, in Brazil. After using S-OPL, the students were asked about benefits and difficulties of using it. As benefits they pointed out that the use of an OPL contributes to the to the quality of the resulting ontology and to the productivity of the development process. As main difficulties, they highlighted: (i) the lack of information about whether, or not, to follow the paths in the OPL process; (ii) difficulties to identify the OPL main flow; (iii) the lack of explicit flow final nodes, indicating where the process finishes; and (iv) problems with variant patterns. In parallel with these studies, we developed the systematic mapping presented in Sect. 2.

Based on the findings of both the two experiments and the mapping, we decided to develop a modeling language for representing OPLs, called OPL-ML. In such endeavor, we started developing a meta-model, as its abstract syntax. Next, we relied on the principles of the Physics of Notations (PoN) [8] for designing a cognitively effective visual notation, and we followed the design process established in the PoN-Systematized (PoN-S) approach [9] to develop its concrete syntax. After developing OPL-ML, we used it to reengineer S-OPL and evaluated it through an experimental study. Aspects regarding the use of PoN-S to develop OPL-ML are discussed in [9], and are out of the scope of this paper. Here, our focus is on OPL-ML, which is presented next.

4   OPL-ML: Ontology Pattern Language Modeling Language

According to [15], a visual language consists of a set of graphical symbols (visual vocabulary), a set of compositional rules for forming valid expressions (visual grammar), and semantic definitions for each symbol (visual semantics). The set of symbols

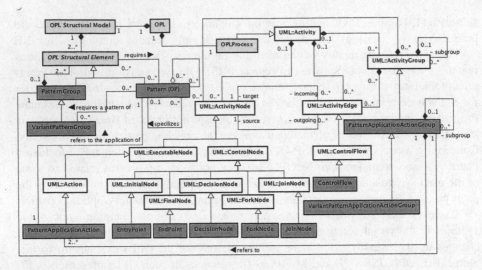

Fig. 1. OPL-ML meta-model.

and compositional rules together form the visual (concrete) syntax. Graphical symbols are used to represent semantic constructs, typically defined by a meta-model. An expression in a visual notation is a diagram. Diagrams are composed of instances of graphical symbols, arranged according to the rules of the visual grammar. In this paper, we focus on the visual syntax of OPL-ML, and the meta-model that defines its constructs, which is shown in Fig. 1.

An OPL comprises both a structural model and a process. The *OPL Structural Model* focuses on patterns and pattern groups. It is composed of *OPL Structural Elements*, which can be of two types: *Pattern* and *Pattern Group*. A *Pattern* represents an OP, while a *Pattern Group* is a way of grouping related OPs and other pattern groups. Thus, a *Pattern Group* is composed of *OPL Structural Elements*. A special type of pattern group is the *Variant Pattern Group*, which is a set of patterns that solve the same problem, but each in a different way. Only one pattern from a *Variant Pattern Group* can be used at a time. *Patterns* that compose a *Variant Pattern Group* are variants of each other. Patterns can be composed of other patterns, and can specialize other patterns. Patterns may also depend on other patterns, i.e., for applying a pattern, another must be applied first. An OPL shall represent dependencies between patterns or between a *Pattern Group* and a *Pattern*. The *requires* relationship captures this dependency. Finally, a *Pattern* may require the application of a pattern from a *Variant Pattern Group*.

Table 1 shows the concrete syntax for representing the elements of the OPL structural model. It is important to say that meta-model elements shown in gray in Fig. 1 do not require a symbol, and thus are not in Table 1. *Patterns* are represented by rectangles, which is the most common symbol used for representing patterns in SE pattern languages. *Pattern Groups* are represented by figures closed by blue straight solid lines (solid polygons). For representing *Variant Pattern Groups*, the same notion was applied, but now using red dashed lines. For allowing managing complexity in large diagrams, a

Table 1. Symbols of the visual notation for OPL structural models.

Element	Symbol
Pattern	<<name>>
Pattern Group (expanded format)	<<name>>
Pattern Group (black box format)	<<name>>
Variant Pattern Group (expanded format)	<<name>>
Variant Pattern Group (black box format)	<<name>>
Relation "requires"	⟶
Relation "requires a pattern of"	------▶
Composition relation	◇—
Specialization relation	⟶▷

black box representation is also provided for pattern groups. These alternative forms are represented by means of rectangles decorated by a rake-style icon (⋔), indicating that this element is detailed in another diagram. This icon is used in UML to represent that an element represented by the decorated construct encapsulates further elements. Finally, regarding relationships, the dependency relations *requires* and *requires a pattern of*, both are represented by an arrow from the dependant to the dependee. For differentiating between them, arrows representing the *requires* association are symbolized with solid lines, in contrast to the dashed lines for the *requires a pattern of* association. This decision is in line with the one of representing *Pattern Groups* using solid lines, and *Variant Pattern Groups* using dashed lines. Pattern composition and pattern specialization are represented by the same symbols used in UML for aggregation and specialization, respectively.

The OPL Process comprises a set of actions devoted to select and apply the patterns, and the control nodes that allow defining a workflow to navigate between the patterns. The part of OPL-ML meta-model describing the OPL Process is an extension of a subset of the UML's meta-model regarding activity diagrams [16], and its concrete syntax is based on the UML notation for activity diagrams. This benefits users who are familiar with this notation. In Fig. 1, classes from the UML's meta-model are shown in white.

An *OPL Process* is a subtype of (UML::) *Activity* specifying a behavior regarding the application of OPs. This behavior is specified as a workflow connecting *Pattern Application Actions* by means of *Control Flows* to other *Pattern Application Actions* and *Control Nodes*. A *Pattern Application Action* is an (UML::) *Action* concerning a pattern application, i.e. this action node refers to the application of a specific OP. For allowing modeling the workflow of the *OPL Process*, the following *Control Nodes* can be used: *Entry Point, End Point, Decision Node, Fork Node*, and *Join Node*.

An *Entry Node* is a control node that acts as a starting point for executing the *OPL Process*. It is a subtype of (UML::) *Initial Node*, and as such, shall not have any incoming *Control Flows*. However, there is an important difference between *Entry Point* and (UML::) *Initial Node*: in a UML activity diagram, if an activity has more than one initial node, then invoking the activity starts multiple concurrent control flows, one for each initial node [OMG, 2015]; in an OPL, if its *OPL Process* has more than one *Entry Point*, this indicates that one and only one of the *Entry Points* should be selected as the starting point. An *End Point*, like its super-type (UML::) *Final Node*, indicates a point at which the workflow stops. Thus, it shall not have outgoing *Control Flows*.

A *Decision Node* is a control node that chooses one between the outgoing *Control Flows*. It is a subtype of (UML::) *Decision Node*, but it presents a slightly different semantics. OPL-ML's *Decision Node* shall have at least one incoming *Control Flow*, and more than one outgoing *Control Flow*. Guard conditions shall be specified to help decide which *Control Flow* to follow. UML's *Decision Node*, in turn, does not admit multiple incoming *Control Flows*. We decided to admit multiple incoming *Control Flows* in OPL-ML to not need to include UML's *Merge Node* in OPL-ML meta-model. As a consequence, the diagrams built using OPL-ML tend to become simpler.

Fork Node and *Join Node* are subtypes of their homonymous counterparts in UML's meta-model, and preserve the same semantics. *Fork Node* is a control node that splits a flow into multiple concurrent flows. It shall have exactly one incoming *Control Flow* and multiple outgoing *Control Flows*, which must be followed. *Join Node* is a control node that synchronizes multiple flows. It shall have exactly one outgoing *Control Flow* but may have multiple incoming *Control Flows*.

Like patterns, *Patterns Application Actions* can be grouped. A *Pattern Application Action Group* groups a set of *Patterns Application Actions*, plus the *Control Flows* and *Control Nodes* establishing the workflow inside the action group. Thus *Pattern Application Action Group* is a subtype of (UML::) *Activity Group*, and as such, may be composed by other *Pattern Application Action Groups*. A *Pattern Application Action Group* should refer to a *Pattern Group* in the OPL Structural Model. When a *Pattern Application Action Group* refers to a *Variant Pattern Group*, then it is said a *Variant Pattern Application Action Group*. A *Variant Pattern Application Action Group* captures alternative *Pattern Application Actions* to be accomplished. Only one of them can be selected. Thus, *Variant Pattern Application Action Groups* do not represent workflows, and they do not admit *Control Flows* and *Control Nodes* inside them.

By developing the OPL Process meta-model as an extension of the UML's meta-model for activity diagrams, we could use the same concrete syntax for most of its elements, namely: Pattern Application Action (UML::Action), Control Flow (UML:: Control Flow), Entry Point (UML::Initial Node), End Point (UML::Final Node), Decision Node (UML::Decision Node), Fork Node (UML::Fork Node), and Join Node (UML::Join Node). See these notations in [OMG, 2015]. Only for groups we decided to make some extensions also in the concrete syntax.

Like in the case of the Structural Model, we decided to provide two ways to represent groups: one as a black box, and the other in an expanded format. In the black box representation, *Pattern Application Action Groups* are represented as UML::*Call*

Table 2. Symbols of the visual notation for groups of pattern application actions.

Element	Symbol
Pattern Application Action Group (black box format)	<< name >>
Pattern Application Action Group (expanded format)	<<name>> <<name>>
Variant Pattern Application Action Group (black box format)	<< name >>
Variant Application Action Group (expanded format)	<< name >>

Behavior Actions with the adornment for calling activities, the rake-style symbol (⋔) [16]. Besides, we suggest setting blue to the border line as a redundant coding [8]. Variant Pattern Application Action Groups are represented by the same symbol, but with red dashed lines. In the expanded format, Pattern Application Action Groups are represented by means of closed regions delimited by blue straight lines, with rounded corners; while Variant Pattern Application Action Groups are represented by means of closed regions delimited by red dashed lines, with rounded corners. Table 2 shows the concrete syntax for representing (Variant) Pattern Application Action Groups.

5 OPL-ML Evaluation

To preliminary evaluate OPL-ML, we used it to reengineer the Service OPL (S-OPL) [14]. Next, we discuss the main changes made during the S-OPL reengineering and present some S-OPL fragments. The full specifications of the first and the current version of S-OPL are available at https://nemo.inf.ufes.br/projects/opl/s-opl/.

(i) *Separation of structural and behavior aspects*: as previously discussed, different models should be used to address structural and behavioral aspects of an OPL. S-OPL was represented only by a process model. Thus, we created a structural model for S-OPL. Figure 2 shows a fragment of the defined structural model.

(ii) *Complexity management*: for managing complexity, we added to S-OPL a general behavior model (Fig. 3) in which patterns groups are presented in the black box format, providing a clearer view of the whole process.

(iii) *Creation of a main mandatory flow:* in the S-OPL old version, the process flow was defined mainly by optional flows (only some flows were mandatory). Thus, when developing an ontology, the ontology engineer could stop following the process at any point (except when there were mandatory flows to follow). According to OPL-ML, a process must be followed from an entry point to an end point. Considering that, we reengineered the S-OPL process creating a main

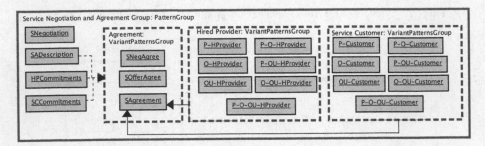

Fig. 2. Structural model of the service negotiation and agreement group.

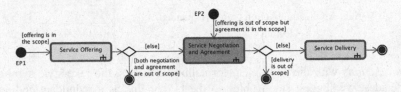

Fig. 3. S-OPL process model (general view).

flow to be followed according to the ontology engineer decisions until s/he reaches an end point.

For creating the main flow, decision nodes were used to indicate optional flows. Decision nodes make it clear that a decision must be taken by the engineer and that one (and only one) of the output flows must be followed. Moreover, we changed the fork node semantics. In the S-OPL old version, outputs of fork nodes were optional flows. According to OPL-ML, outputs of fork nodes are mandatory flows, i.e., when following the input flow of a fork node, all its output flows must be followed. In order to not interrupt the process main flow, we used joint nodes to converge the output flows into a single one that take up the main flow. Finally, aiming to improve the process flow, we made some changes in patterns grouping. Figure 4 shows a fragment of the

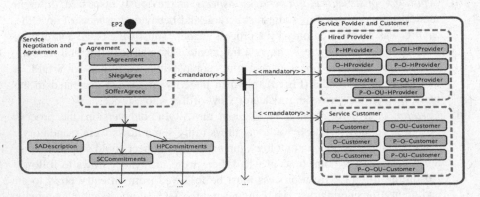

Fig. 4. Fragment of the old version of S-OPL process model.

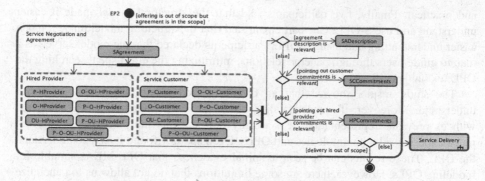

Fig. 5. Fragment of the reengineered S-OPL process model.

old version of the process model defined for the Service Negotiation and Agreement group. Figure 5 presents the corresponding reengineered model.

After reengineering S-OPL, aiming to evaluate its new version, we performed a study involving 6 of the 9 students that participated in the previous study with S-OPL. The study goal was to evaluate: (i) if an OPL represented by using OPL-ML is easier to understand, and (ii) if the S-OPL new version is better than the previous one. The study procedure consisted of two steps. In the first step, the participants received the new specification of S-OPL and analyzed it. The specification contained the process and structural models, their descriptions, and the documentation of each pattern (the same of the previous version). In the second step, they provided feedback about the new version of S-OPL by filling a questionnaire containing two parts. The first one included questions about the difficulty for understanding the OPL, especially the different models representing structural and process aspects, and each one of the OPL symbols. The possible answers used a Likert Scale: very hard, hard, neutral, easy, and very easy. The second part concerned comparing the S-OPL new version with the previous one, considering process clarity and ease of understanding, and support provided by the structural model. Another questionnaire was applied to capture the participants' profile, analyzing their level of education, experience in conceptual modeling, experience in ontology development, and experience with OPLs. The group of participants has, mostly, medium experience (from one to three years) in conceptual modeling, medium/low experience in ontologies development and low experience (less than one year) with OPLs.

Concerning the OPL mechanism, one participant found it very easy, four found it easy, and one was neutral. As for the OPL symbols, except by one participant, who declared that it was difficult to understand variant patterns, all found easy or very easy to understand the symbols and reported that the notation is clear.

Comparing the S-OPL new version with the previous one, five participants said that in the new version the process is much clearer and one participant said that it is clearer. The participants commented that the main flow defined in the process made it easier to understand the process and follow it. Moreover, all participants found the new version better than the previous one. They pointed out that the different abstraction levels contribute to better understand the whole process, turning the process more intuitive and practical. Finally, five participants said that the structural model made it easier understand and use S-OPL. One participant said that the structural model made it much easier understanding and using S-OPL. Participants declared that the model served as a map to guide users through patterns relations, minimizing the effort spent searching the OPL textual specification.

The study results showed that an OPL modeled using OPL-ML is easier to understand. Moreover, the understanding about the OPL improved when compared with its previous representation. It is worthy noticing that the participants were not much experienced with ontologies and OPLs. Even so, they found it easy to understand the OPL. These results can be seen as initial evidences that OPL-ML is suitable for modeling OPLs. However, there are some limitations that do not allow us to generalize the results. As main limitations, we can point out: (i) the small number of participants in the study; (ii) the study was performed in an academic environment; (iii) when evaluating the new version of S-OPL, the participants had already used the S-OPL previous version, thus, the acquired knowledge about S-OPL might have influenced the understanding about its new version.

6 Related Works

As previously discussed, the use of OPLs is recent and the few works defining OPLs were done by the same research group, the Ontology and Conceptual Modeling Research Group (NEMO). This research group has proposed five OPLs [7]. In these OPLs, extensions of UML activity diagrams are used to represent the OPL processes. When comparing OPL-ML with the visual notation described in [7], the main similarity is that both represent the OPL process by means of extensions to the UML's activity diagram. As for differences, several advances are introduced by OPL-UML.

First, the visual notation described in [7] is limited, being not able to cover all aspects for properly representing an OPL. OPL-ML clearly defines the language abstract and concrete syntaxes. The abstract syntax is defined by a meta-model and part of it extends the UML meta-model for activity diagrams, including constructs not considered in [7] (e.g., end point). As for the concrete syntax, it was defined by following a systematic process applying the PoN principles, resulting in a cognitively rich visual notation.

Second, the OPLs presented in [7] are represented only by a process model. Structural aspects are not addressed, being difficult for users to identify relations between the patterns. OPL-ML proposes the use of two different models for representing an OPL: the structural model, dealing with the OPL structural elements and the relationships between them; and the process model, addressing the process to be followed for selecting and applying the patterns. Separating these views in two types of models contributes to better understand the OPL and the relationships between its elements.

Third, OPL-ML allows managing complexity by using models at different abstraction levels. A general model, using black box formats, can be created to represent a whole view of the OPL, while detailed models expand the black boxes and provide a detailed view of the OPL. By using this resource, large or complex OPLs can be easier understood. The visual notation described in [7] does not provide this facility.

Finally, by using OPL-ML, a main flow is defined in the OPL process, making it clear the paths (from beginning to end) to be followed by the ontology engineer according to his/her decisions. In the OPLs presented in [7], the process does not have a continuous flow and can be interrupted at any time.

As discussed in Sect. 2, the use of pattern languages (PLs) in Software Engineering (SE) are much more mature than in Ontology Engineering. This fact motivated us to use insight from SE PLs to define OPL-ML. However, despite of the use of PLs is already consolidated in SE, there are still problems regarding their visual representation. For instance, considering the investigated software-related PLs, only two of them use different models to address structural and behavioral aspects. In [17], Guerra et al. propose a PL for organizing the internal structure of metadata-based frameworks, which is represented by structural and navigation models. In [18], in turn, Zdun proposes a PL for designing aspect languages and aspect composition frameworks. Patterns and relationships between them are addressed by a relationship model, while a feature model helps patterns selection according to the features to be considered when designing frameworks. Concerning the structural model, Guerra et al. [17] address only dependency relationships, while in [18] the types of relationships vary and are not clear. As for the process model, the navigation model described in [17] presents several sequences in which patterns can be applied. However, the visual notation is limited, being not possible, for example, to identify where the process starts or ends and which paths are mandatory. The feature model presented in [18], although can be helpful to patterns selection, does not provide a process to be followed. Thus, the sequence in which the patterns should be applied is not clear.

7 Final Considerations

As pointed by Blomqvist et al. [3], the works already done aiming to provide Ontology Pattern Languages (OPLs) are the first steps towards developing an ontology pattern representation language. In this paper, we take a step forward by providing a modeling language for representing OPLs: OPL-ML. Such language presents as striking features the following: (i) OPL-ML defines explicitly its abstract syntax by means of a meta-model; (ii) OPL-ML concrete syntax is designed following Moody's principles for designing cognitively effective visual notations [8]; (iii) OPL-ML is designed considering the results of a systematic mapping on visual notations for representing Software Engineering pattern languages, and from experimental studies involving the use of OPLs. We believe that the definition of a visual language for modeling OPLs amplifies the available resources for ontology engineers to develop OPLs, increasing productivity and contributing for advances in the area.

OPL-ML was used to reengineer S-OPL, a service OPL proposed in [14]. With S-OPL new version in hands, we performed a study to get perceptions from the OPL users. The findings indicate that OPL-ML use is viable and that it is capable of properly representing OPLs. However, the evaluation carried out by now is limited. Thus, new studies must be performed aiming to better evaluate OPL-ML. As future studies, we intend to reengineer the OPLs already developed at NEMO, develop new OPLs by using OPL-ML and evaluate the use of OPL-ML by other people, who can provide new feedback regarding OPL-ML use and effectiveness.

Besides, we have been working on the development of a supporting tool to aid ontology engineers in creating OPLs and using them to build ontologies.

Finally, although OPL-ML has been proposed to represent OPLs, we believe that it can be adapted and used to represent pattern languages in general. In this sense, we intend to investigate how to generalize OPL-ML, maybe including other elements typically used in other Software Engineering pattern languages.

Acknowledgements. This research is funded by the Brazilian Research Funding Agency CNPq (Processes 485368/2013-7 and 461777/2014-2) and FAPES (Process 69382549/2014).

References

1. Poveda-Villalón, M., Suárez-Figueroa, M.C., Gómez-Pérez, A.: Reusing ontology design patterns in a context ontology network. In: Second Workshop on Ontology Patterns, WOP 2010, Shangai, China (2010)
2. Falbo, R.A., Guizzardi, G., Gangemi, A., Presutti, V.: Ontology patterns: clarifying concepts and terminology. In: Proceedings of the 4th International Conference on Ontology and Semantic Web Patterns, vol. 1188, pp. 14–26. CEUR-WS.org (2013)
3. Blomqvist, E., Hitzler, P., Janowicz, K., Krisnadhi, A., Narock, T., Solanki, M.: Considerations regarding ontology design patterns. Semant. Web 7(1), 1–7 (2015)
4. Ruy, F., Guizzardi, G., Falbo, R.A., Reginato, C.C., Santos, V.A.: From reference ontologies to ontology patterns and back. Data Knowl. Eng. **109**, 41–69 (2017)
5. Blomqvist, E., Gangemi, A., Presutti, V.: Experiments on pattern-based ontology design. In: Proceedings of K-CAP 2009, pp. 41–48 (2009)
6. de Almeida Falbo, R., Barcellos, M.P., Nardi, J.C., Guizzardi, G.: Organizing ontology design patterns as ontology pattern languages. In: Cimiano, P., Corcho, O., Presutti, V., Hollink, L., Rudolph, S. (eds.) ESWC 2013. LNCS, vol. 7882, pp. 61–75. Springer, Heidelberg (2013). doi:10.1007/978-3-642-38288-8_5
7. Falbo, R.A., Barcellos, M.P., Ruy, F.B., Guizzardi, G., Guizzardi, R.S.S.: Ontology pattern languages. In: Gangemi, A., Hitzler, P., Janowicz, K., Krisnadhi, A., Presutti, V. (eds.) Ontology Engineering with Ontology Design Patterns: Foundations and Applications. IOS Press (2016)
8. Moody, D.L.: The "Physics" of notations: toward a scientific basis for constructing visual notations in software engineering. IEEE Trans. Softw. Eng. **35**(6), 1–22 (2009)
9. da Silva Teixeira, M.G., Quirino, G.K., Gailly, F., de Almeida Falbo, R., Guizzardi, G., Perini Barcellos, M.: PoN-S: a systematic approach for applying the physics of notation (PoN). In: Schmidt, R., Guédria, W., Bider, I., Guerreiro, S. (eds.) BPMDS/EMMSAD - 2016. LNBIP, vol. 248, pp. 432–447. Springer, Cham (2016). doi:10.1007/978-3-319-39429-9_27
10. Wieringa, R.J.: Design Science Methodology for Information Systems and Software Engineering. Springer, Heidelberg (2014)
11. Schmidt, D., Stal, M., Rohnert, H., Buschmann, F.: Pattern-Oriented Software Architecture, Volume 2: Patterns for Concurrent and Networked Objects. Wiley Publishing, New York (2000)

12. Kitchenham, B., Charters, S.: Guidelines for performing systematic literature reviews in software engineering - Version 2.3. EBSE Technical report EBSE -2007-01 (2007)
13. Ruy, F.B., Falbo, R.A., Barcellos, M.P., Guizzardi, G., Quirino, G.K.S.: An ISO-based software process ontology pattern language and its application for harmonizing standards. SIGAPP Appl. Comput. Rev. **15**(2), 27–40 (2015)
14. Falbo, R.A., Quirino, G.K., Nardi, J., Barcellos, M.P., Guizzardi, G., Guarino, N., Longo, A., Livieri, B.: An ontology pattern language for service modeling. In: 31st ACM/SIGAPP Symposium on Applied Computing, Pisa, Italy (2016)
15. Moody, D.L., Heymans, P., Matulevicius, R.: Visual syntax does matter: improving the cognitive effectiveness of the i* visual notation. Requir. Eng. **15**(2), 141–175 (2010)
16. OMG. OMG Unified Modeling Language Version 2.5 (2015)
17. Guerra, E., Alvez, F., Kulesza, U., Fernandes, C.: A reference architecture for organizing the internal structure of metadata-based frameworks. J. Syst. Softw. **86**(5), 1239–1256 (2013)
18. Zdun, U.: Pattern language for the design of aspect languages and aspect composition frameworks. IEE Proc. Softw. **151**(2), 67–83 (2004)

QMMQ 2017 - 4th Workshop on Quality of Models and Models of Quality

Preface

The 4th workshop on Quality of Models and Models of Quality (QMMQ 2017) is hold in Valencia (Spain). This workshop aims to provide an opportunity for researchers and industry developers working on various aspects of information systems quality to exchange research ideas and results and to discuss them. Moreover, QMMQ aims to promote research on information systems and conceptual modeling quality to the broader conceptual modeling research community attending ER 2017. This year QMMQ has a special focus on the quality of business processes, even though the presented papers cover a broad range of topics related with the quality in conceptual models.

We have reviewed a total of 9 submissions of which 4 have been accepted as full papers for the workshop. Submitted papers come from different countries of several continents, such as Chile, Colombia, France, Italy, Spain, Germany and Slovakia. We would like to deeply thank all the authors for submitting their contribution to QMMQ 2017 and are pleased to express our appreciation to all Program Committee for their involvement in the evaluation of submitted papers. Without their expertise and dedication the QMMQ program would not be of such high quality. We would like also thank the co-chairs of the ER workshops since they have helped us to coordinate QMMQ 2017 within the ER 2017 conference.

Finally, we are glad to welcoming you to Valencia. We hope you enjoy this workshops and the benefit from the scientific program but also from sharing your experiences with colleagues from all around the world.

November 2017

Samira Si-Said
Ignacio Panach
Pnina Soffer

Evaluating Quality Issues in BPMN Models by Extending a *Technical Debt* Software Platform

Fáber D. Giraldo(✉) and Fabián D. Osorio

SINFOCI Research Group, University of Quindío, Armenia, Colombia
{fdgiraldo,fdosorios}@uniquindio.edu.co

Abstract. Taking into account the current role of modelling at organizational levels, the *quality* of business process models (i.e., models that indicate how to achieve business objectives) is an essential aspect for the development and technological support of any organization. Based on the hypothesis that the quality of business process models in the BPMN notation can be automatically analyzed, in this work we propose an extension of a code-quality software platform (the SonarQube project) through a complement that allows the quality of these models to be validated. This paper reports the guidelines and quality metrics that were used to evaluate BPMN models. Then the SonarQube code evaluation platform and a plugin that was created in this work (which contains the identified metrics) was used to automatically obtain values of the *technical debt* from BPMN models.

Keywords: Model-driven engineering · Technical debt · BPMN models · SonarQube · Quality rules

1 Introduction

Currently, organizations choose the BPMN notation for modelling their business processes following a workflow format, which is understandable for the involved stakeholders. Business modelers often focus on describing the process without taking into account their quality. Verifying the quality of business process models is a manual task that is influenced by subjective criteria in accordance with the experience and knowledge of the reviewers. There is no predefined, standard qualitative context for models of this type.

The number of research works about quality assurance in BPMN models has being increasing in the last few years. For example, the authors in [1] present a general vision of criteria and measures for quantifying the different aspects of quality in business process models. In [4], the author shows how a quality analysis on BPMN models can be performed by using the XML plugin[1] of the

---

[1] https://docs.sonarqube.org/display/PLUG/SonarXML.

© Springer International Publishing AG 2017
S. de Cesare and U. Frank (Eds.): ER 2017 Workshops, LNCS 10651, pp. 205–215, 2017.
https://doi.org/10.1007/978-3-319-70625-2_19

SonarQube[2] project and the BPMN 2.0 schema that is provided by OMG[3]. This work (called *BPMNspector*[4]) is a starting point for demonstrating the feasibility of the operationalization of the quality analysis of business processes models under the BPMN notation. This strategy evaluates individual files or directories with BPMN files and reports the non-compliance of BPMN 2.0 constraints. However, this analysis is restricted to specific XML rules under the BPMN 2.0 schema, which is equivalent to a static analysis at the lexical level of these models.

Modelling business processes with *good quality* is not an easy task. For this reason, in 2015 we proposed the automatic evaluation of quality in models by integrating model-driven engineering (MDE) projects and approaches for calculating the *Technical Debt* (TD), which traditionally focus on source code from programming languages. The TD term is a metaphor that was introduced by agile practitioners as a way of describing the long-term cost and consequences associated with optimal software design and implementation. Our approach is based on the calculus of TD directly on models that are instantiated from modelling languages (i.e., business process models that are specific BPMN notation). Therefore, the rules that are required to define and analyze technical debt issues are specified at the level of the modelling languages instead of source code.

Our strategy involves the definition of quality rules, which are extracted from the literature about metrics in BPMN models, and the implementation of these quality rules for the SonarQube platform. The *SonarQube* open-source platform was chosen because of its support of the TD calculation and continuous inspection and because of the capacities that are provided for extending (and including) new rules for assessing TD. Other previous works report the use of SonarQube in the assessment of models. An example is [7], in which the authors report a SonarQube plugin for assessing Event-Driven Process Chains models using information that is generated from business modelling tools.

The remainder of this article is structured as follows: Sect. 2 describes the main sources of metrics for BPMN models that were chosen from the literature. Section 3 presents the design and implementation of the Eclipse-based plugin and the quality rules for BPMN in the SonarQube platform. Section 4 presents the preliminary validation procedures that were performed for the project. Finally, conclusions and further work are presented.

2   Sources of Quality Metrics for BPMN Models

In the following, we present two sources of metrics for the BPMN models that were used in this work:

2.1   7PMG

The 7PMG framework [5] defines a set of metrics for guiding the improvement of quality in BPMN models so that these models can be understandable for all

---

[2] https://www.sonarqube.org/.
[3] http://www.omg.org/spec/BPMN/2.0/.
[4] https://github.com/uniba-dsg/BPMNspector.

of the stakeholders involved. 7PMG provides a set of recommendations that can be used to build a process model from scratch or to improve already existing models. These guidelines are based on the idea that there are different ways of describing a behavior using a process model. Therefore, it is possible to find the same business process modelled in different ways that comply with specific quality guidelines without being described in the same way.

The 7PMG guidelines promote the understandability of process models; therefore, one example of guidelines is the use of a maximum number of elements in one business process model canvas due to the possible cognitive effects by large diagrams (e.g., a low level of understandability by users). Other 7PMG guidelines that we considered for our SonarQube plugin are presented in Table 1.

Table 1. The 7PMG guidelines that were used in the implementation of our SonarQube plugin.

Guideline	Description	Additional considerations
G - 1	The model must use as low a number of elements as possible. The size of a model directly affects the understandability of the model and increases the probability of mistakes in the model	Models with a higher number of elements tend to be more difficult to understand; thus, impeding the goal of understandability for all of the stakeholders involved
G - 2	The modeler must establish the lowest number of relationships for each element	Each element of the diagram has a weight; it refers to the number of relations that arrive (incoming arcs) and/or leave (outgoing arcs) it
G - 3	Only a start event and an end event can be used to build a model. Models that meet this guideline are more suitable for analytic purposes	It is allowed to use a starting event and an ending event for each pool of a BPMN diagram
G - 5	Avoid the use of inclusive gateways. The semantics of these elements produces paradoxes and problems in the implementation of the model	BPMN models with exclusive and parallel gateways tend to have a lower number of mistakes
G - 7	The model must be broken down if it has more than fifty elements (G-1 guideline)	Models with more than fifty elements increase the probability of mistakes by 50%. These models should be split into small models. Another alternative is to restructure these models by using the sub-process element of BPMN

2.2 Specialization of the SEQUAL Framework for BPMN Models

A set of guidelines for BPMN models is proposed in [2]. These guidelines are a specialization of the SEQUAL framework, one of the most relevant initiatives for evaluating quality at the model-based level [3]. Guidelines are structured in

Table 2. SEQUAL specialized guidelines for BPMN that were used in the implementation of our plugin.

Guideline	Description	Additional considerations
NE - 1	The model contains a high number of elements	Models with more than 31 elements must be split
NE - 5	The model contains several starting/ending events.	Avoid models with more than two starting/ending events at the superior level of the models, using a starting event for sub-processes and two ending events to distinguish the *successful* and *failed* states
NE - 6	Do not avoid the starting event and the ending event	The model must contain at least one starting event and one ending event
NE - 8	A high number of *arcs* in the model	Avoid models with more than 34 arcs
NE - 9	A high number of *gateways* in the model	Avoid models with more than 12 gateways
NE - 13	A high number of outgoing sequence flows from an event	Avoid 4 or more outgoing sequence flows in an event
CC - 1	The model has internal structured blocks	Avoid models with internal structured blocks
CC - 2	The model has several cycles	Avoid cycles with multiple output points
CC - 5	The sum of the output degrees of *AND* and *XOR* gateways should be less than or equal to 8	Avoid a high level of parallelism in BPMN models
CC - 10	There are inclusive gateways in the model	Avoid inclusive gateways
CC - 12	The model does not have modularity	A model should not have more than 31 elements in a diagram. A sub-process should not have less than 5 activities
P - 1	Maintain the diagram in an ordered and consistent form	Minimize the number of lines that needlessly cross each other and the number of overlapped elements (i.e., borders should not overlap other borders or nodes)

three groups: *number of different elements in the model, composition of components*, and *presentation*. Table 2[5] presents the guidelines from this SEQUAL' specialization that were implemented in our plugin.

---

[5] NE: Number of Elements, CC: Composition of Components, P: Presentation.

3   The SonarQube Plugin Implemented for the Calculation of TD in BPMN Models

From the guidelines presented in Tables 1 and 2, we defined quality rules that were implemented in our plugin for the SonarQube platform. The incorporation of these rules in SonarQube was made by programming every rule in the Java language. In addition, a plugin that receives a BPMN model in XML format (XMI) as input was created.

This plugin is incorporated into SonarQube so that when the model is validated, it shows the results of the mistakes found in the model. A SonarQube plugin is composed of Java classes that implement *extension points* (i.e., interfaces) and an abstract class that defines the methods to be implemented. A plugin has rules, which are classes that read a file and report the issues that are found in the file [6]. Plugins are the mechanism used to extend TD analysis in SonarQube.

In addition, a TD value must be established for each rule. This TD value contains the associated time and costs for fixing the detected quality issue. For demonstration purposes, we defined some preliminary time values in our plugin. Table 3 shows the implemented rules with their TD values. We are planning further work to precisely determine TD values for quality rules of BPMN models.

Each implemented rule is produced by the operationalization of the guidelines presented in Sect. 2. For example, the *Number of Elements in a Diagram* rule was defined from the *G-1* and *G-7* guidelines from the 7PMG framework and from the *N-1* and *CC-12* guidelines from the specialized SEQUAL. Each rule has a Java block code associated to it. These codes were later added to a SonarQube server instance.

To demonstrate the technical feasibility of our approach, we made a BPMN model in Bonitasoft (Fig. 1) from an example that we found on the official site of this tool (a Helpdesk model[6]). Figure 2 presents the results of the analysis reported by the SonarQube platform for the BPMN model in Fig. 1. The results show the number of quality issues detected in the BPMN model, which are according to their level of importance (i.e., *Blocker*, *Critical*, *Major*, *Minor*, and *Info*) following the SonarQube conventions. Figure 3 shows the report of a quality issue directly in the XMI code of the BPMN model. Finally, Fig. 4 presents an example of a description of the rule for a chosen quality issue.

---

[6] Example available in http://www.bonitasoft.com/for-you-to-read/process-library/it-help-desk.

Table 3. The rules implemented in the BPMN plugin for SonarQube.

Rule	Purpose	Associated *Technical debt* value
Number of elements in a diagram	To count the number of elements in a canvas	Four minutes are added for each extra element in the diagram
Number of connections for each element	To count the number of connections that each element in the diagram has	Six minutes are added for each extra connection
Number of connections in a diagram	To count the number of connections that a BPMN model has. It is useful to determine the *weight* of each element in the diagram	Twenty minutes are added if the total weight exceeds the maximum value for it
Number of incoming arcs	To count the number of incoming arcs or relations for a specific element	Four minutes are added for each extra arc
Number of outgoing arcs	To count the number of outgoing arcs or relations for a specific element	Eight minutes are added for each extra arc
Degree of an element	To determine the *degree* of an element in the diagram, i.e., the number of incoming and outgoing connections in an element	Twenty minutes are added if the *degree* value exceeds the maximum value for it
Number of OR gateways that are used in the diagram	To count the number of OR gateways in the diagram. BPMN models that frequently use these gateways increase the probability of mistakes	A *warning* indication is presented to indicate the probability of inserting mistakes
Existing elements in each Pool	To count the number of elements in each Pool of the diagram. This rule verifies that not all elements in the model are in only one Pool	Twelve minutes are added for each Pool that exceeds the maximum number of elements. In addition, a warning message is presented in the report to suggest the split of the number of elements
Task in a lane	To identify multiple instances of a same task that represents multiple participants	If a lane contains only one element and this element is an instanced task, a warning indication is generated to suggest the suppression of the repeated task. Eight minutes are added
Number of starting events	To count the number of starting events in a model	Three minutes are added for each extra start event in the model
Number of ending events	To count the number of ending events in a model	Three minutes are added for each extra end event in the model

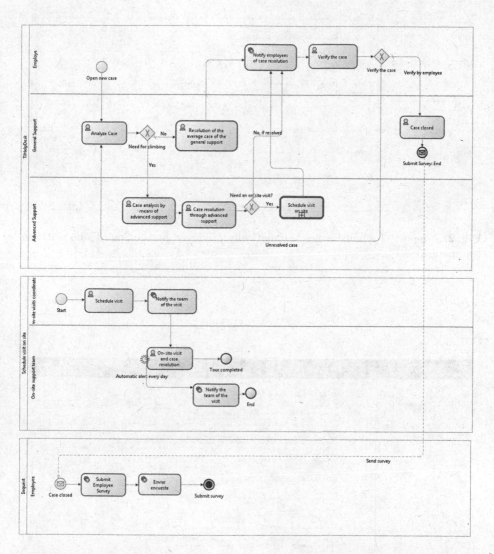

Fig. 1. The BPMN model used to demonstrate the applicability of quality rules.

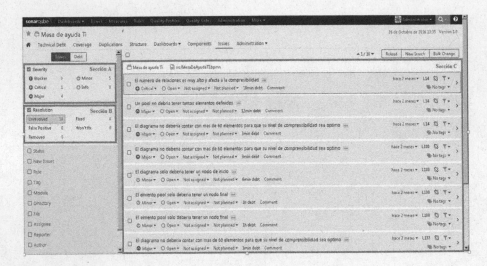

Fig. 2. A report of quality issues automatically produced by the SonarQube platform (in Spanish).

Fig. 3. A report of quality issues in XMI code of the BPMN model.

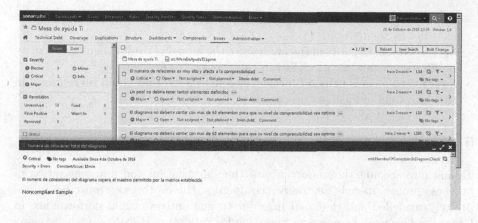

Fig. 4. An example of a report of the *Degree of an element* rule (in Spanish).

4   A Preliminary Validation of Our Approach Through Expert Judgment

Taking advantage of the tutorial that was presented in the International Seminar of Computing Science (SICC 2016)[7] in Colombia, an expert judgment procedure with experts from France and Chile was performed.

The researchers demonstrated the main features of the project, and the experts reviewed the proposal and performed the process of validation of the quality of a BPMN model in SonarQube using the Bonitasoft editor[8]. The experts provided feedback, so these proposals are planned as future works that are derived from our approach. Among the proposals given by the experts are the following:

– *An interest in external metrics*, i.e., calculating the interest that each debt generates for the purpose of validating not only the absolute debt, but also the interest, which is generated when the issue is not addressed on time.
– *Internal variables of the technical debt*, i.e., reviewing and adjusting the internal variables that SonarQube uses to validate the technical debt.
– Taking the values of the resulting TD to determine the feasibility of reaching a business goal, i.e., whether the business goal that is associated to a business process model is viable, or, at least, could comply with the process.
– Verifying the selected guidelines with regard to the context of the business process model. There are many ways to represent a process, although the model with the lowest number of elements is not always viable.
– Verifying the *subjectivity* of a BPMN model by validating it with respect to the business requirements or business rules that belong to the organization in which the BPMN models are proposed.

---

[7] Information available in https://sites.google.com/site/sicc2016medellin/.
[8] Available in http://www.bonitasoft.com/.

- Exploring possibilities of managing organizational rules for modelling as quality rules in SonarQube.
- *Secure Tropos*: to perform the analysis of the model of complementary security requirements that are modelled in the Secure Tropos language[9].
- Integrating models of non-functional requirements that could be required in the modelling of business processes.

5   Conclusions

Taking into account the importance of the specification and automatization of business process models for today's companies, these processes must be appropriately modelled and designed in order to guarantee optimal performance in later use. This work presents an approach for validating BPMN models by using (and extending) the SonarQube platform in order to find quality issues that are associated to the models and to estimate their associated technical debt.

Eleven rules for BPMN models were implemented as a plugin for the SonarQube platform. These rules were based on two important quality guidelines that were previously reported in the literature. The SonarQube platform allows BPMN models to be analyzed at the lexical, syntactic, and (partially) semantic levels by using our implemented plugin.

This allows analyzing BPMN models in early phases of the implementation of these type of models, avoiding later consequences in projects driven by business process models, such as the deployment of processes and misalignment between business goals and resulting business processes models. In addition, quality dimensions such as the understandability of business process models can be determined in order to support all of the stakeholders that are involved in a modelling project.

Acknowledgements. This work has been supported by the University of Quindío (Colombia) through the Research Project 742 - *Integración conceptual, metodológica y tecnológica del concepto de Deuda Técnica en entornos Model-Driven Engineering soportados por CIAT* - funded by the Vice-chancellor's Research Office (grant 2015–02).

References

1. Kahloun, F., Channouchi, S.A.: Quality criteria and metrics for business process models in higher education domain: case of a tracking of curriculum offers process. Procedia Comput. Sci. **100**, 1016–1023 (2016). International Conference on ENTERprise Information Systems/International Conference on Project MANagement/International Conference on Health and Social Care Information Systems and Technologies, CENTERIS/ProjMAN / HCist 2016
2. Krogstie, J.: Quality in Business Process Modeling. Springer, Cham (2016). ISBN 978-3-319-42510-8

---

[9] http://www.troposproject.org/node/301, http://austria.omilab.org/psm/content/sectro/info.

3. Krogstie, J.: SEQUAL Specialized for Business Process Models, pp. 103–138. Springer International Publishing, Cham (2016). ISBN 978-3-319-42512-2
4. Geiger, M.: Sonarqube tutorial: Xml-plugin - setup and usage (matthias geiger's blog), March 2014
5. Mendling, J., Reijers, H.A., van der Aalst, W.M.P.: Seven process modeling guidelines (7pmg). Inf. Softw. Technol. **52**(2), 127–136 (2010)
6. Solovjev, A.: Sonar plugin development, January 2013
7. Storch, A., Laue, R., Gruhn, V.: Measuring and visualising the quality of models. In: 2013 IEEE 1st International Workshop on Communicating Business Process and Software Models Quality, Understandability, and Maintainability (CPSM), pp. 1–8 (2014)

Utility-Driven Data Management for Data-Intensive Applications in Fog Environments

Cinzia Cappiello, Barbara Pernici, Pierluigi Plebani, and Monica Vitali[✉]

DEIB, Politecnico di Milano, Piazza Leonardo da Vinci 32, 20133 Milano, Italy
{cinzia.cappiello,barbara.pernici,pierluigi.plebani,
monica.vitali}@polimi.it

Abstract. The usage of sensors, smart devices, and wearables is becoming more and more common, and the amount of data they are able to generate can create a real value only if such data are properly analyzed. To this aim, the design of data-intensive applications needs to find a balance between the value of the output of the data analysis – that depends on the quality and quantity of available data – and the performance.

The goal of this paper is to propose a "data utility" model to evaluate the importance of data with respect to their usage in a data-intensive application running in a Fog environment. This implies that the data, as well as the data processing, could reside both on Cloud resources and on devices at the edge of the network. On this basis, the proposed data utility model puts the basis to decide if and how data and computation movements from the edge to the Cloud – and vice versa – can be enacted to improve the efficiency and the effectiveness of applications.

Keywords: Data utility · Cloud and Fog computing

1 Introduction

With an increasing trend, data-intensive applications are becoming fundamental for the analysis of data gathered by the Internet of Things (IoT) [17]. In fact, data collected through tiny and affordable sensors, and transmitted with smart devices, are enabling the fourth industrial revolution supporting, for instance, predictive maintenance of machineries, real-time tracking of production lines, as well as efficient scheduling of tasks. At the same time, mobile phones and wearables are changing the habits of people, as the data collected by these devices can be exploited to optimize the daily activities and improve quality of life.

As the available amount of data increases, data-intensive applications require more and more resources to properly manage and process such data. This is witnessed by the numerous tools available to support both data transmission and data processing, where scalability is the key feature (e.g., Apache Kafka, Apache Spark, and Apache Flume). Regardless of the specific solution, efficiency in data processing is ensured by specific file systems (e.g., HDFS) enabling a proper data

© Springer International Publishing AG 2017
S. de Cesare and U. Frank (Eds.): ER 2017 Workshops, LNCS 10651, pp. 216–226, 2017.
https://doi.org/10.1007/978-3-319-70625-2_20

management: data are spread among different nodes to enable parallel computation, replication is allowed for improving the reliability, and data formats adopt grammars enabling efficient parsing. Furthermore, the computation usually relies on resources available on the Cloud, implying the possibility to easily scale in/out the application with respect to the amount of data to be processed.

The goal of this paper is to propose a *data utility model* evaluating the usefulness of data for an application with respect to the content of the data source, and also to the quality of the content and the data source location (on the Cloud or on the Edge), which could affect the quality of data provisioning. In this paper we present the first step towards this direction: the definition of a model including all the components affecting the data utility when application is running on a Fog environment.

The rest of the paper is organized as follows. To better clarify our approach, we use the running example introduced in Sect. 2. Section 3 discusses the conceptual model of data-intensive applications running in a Fog environment. Section 4 introduces the data utility model for the previously defined data-intensive applications. Finally, Sect. 5 gives an overview of the state of the art in this field, whereas Sect. 6 concludes the paper outlining possible future work.

2   Running Example

Figure 1 draws a possible scenario, in the ambient intelligence domain, that is used along this paper. The data-intensive application to be developed analyzes the comfort in a building using the data coming from sensors placed in different rooms (e.g., temperature, humidity, and brightness sensors). The application also uses weather data, made available by external entities, to perform a validation of the sensors and to predict possible variations suggesting actions to the users through a dashboard. The tasks composing the application are: (i) *Ambient Sensing Alignment (Task A)*: it collects data coming from a set of sensors placed in the monitored building and performs some pre-processing operations, such as timestamps alignment and data cleaning. (ii) *Ambient Sensing Aggregation (Task B)*: it uses the output of the previous step to perform statistical analysis and aggregations (minimum, maximum, and average values for each sensor or for several sensors of the same kind in the building) which constitute the data set relevant for the analysis. (iii) *Data Enrichment and Prediction (Task C)*: it integrates data produced by the previous task with information about weather in the city where the monitored building is placed. (iv) *Visualization Preparation (Task D)*: it prepares the information obtained from the analysis for providing visualization tools to the final user.

Each task has different requirements in terms of data sources to be accessed. Input of a task can be the output of another task ($E_{i,j}$ representing the exchanged information from task t_i to task t_j) or a data sources that could be either directly managed by the application developer or provided by external entities. Orthogonally, data sources can be placed on devices at the edge of the network or they reside in the Cloud. In our example, *Task A* will access to data generated by a well-defined set of IoT devices (i.e., placed at the edge of the network), producing streaming and real time information about the monitored building (Building Sensors DS_B in Fig. 1). *Task C* uses weather data sets placed in the Cloud (Weather Data DS_W in Fig. 1). While for *Task A* a specific data source is defined through the identification of the building to be monitored, for *Task C* we assume that several public data sources fit the requirements of the application. Deciding which is the best data source to be used and where to place the tasks or to move the data (i.e., at the edge or in the cloud) impacts the efficiency and the effectiveness of the application. For instance, executing the *Task A* and *Task B* on a device close to the sensors, rather than on the Cloud, could reduce the amount of data transmitted over the network and thus improve performance. At the same time, if the device on which we aim to execute those tasks has a low performance, it may become a bottleneck for the application. Moreover, the different weather data sources could have different quality levels, so the selection could finally affect the quality of the analysis in *Task C*.

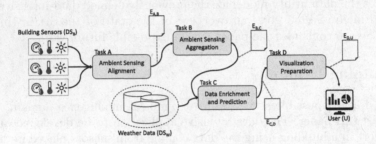

Fig. 1. Example of Data-intensive application in ambient intelligence domain

To make the designer mainly focused only on the application logic, in our approach we introduce the Data Utility as a complex metric which reflects to which extent a data source satisfies the requirements of an application. With the definition of the Data Utility we want to reduce the burden, from the developer standpoint, of selecting the data sources and the location of the tasks.

3   Data-Intensive Applications in Fog Environment

Data-intensive applications are mainly defined by the data to be processed and the computation performed on them. To this aim, a data-intensive application is often modeled with a data flow that defines the data sources that feed the application, as well as the steps to be performed to acquire, manage, and transform such data. The literature already proposes some approaches for modeling data-intensive applications. For instance, in [7] a UML profile has been specifically designed to capture the dependencies between the tasks operating on the data and the data themselves. Yet, the meta-models in [14] cover a broader spectrum: from the business view to a more technical view.

A peculiar aspect of our approach relies on the adoption of Fog computing paradigm for designing and running the application. According to the definition in [15], Fog computing builds upon the capabilities of Cloud computing, extending them toward the edge of the network. As a consequence, data-intensive applications can consider Cloud and Edge as a continuum where both data and tasks can be moved from the edge to the Cloud and vice versa. The designer, instead of specifying a precise deployment plan, specifies the characteristics that a node should, or must, have to run a task or to store data. At execution time, the deployment is adapted, while the application is running, to improve the efficiency and the effectiveness of the application. Focusing on the design standpoint, a *Data Intensive Application* designed to run on a Fog environment relies on three main elements: resources, data sources, and tasks (Fig. 2).

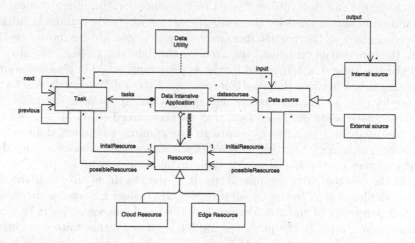

Fig. 2. Data-intensive application model.

Resource. Adopting the Fog Computing paradigm, the infrastructural layer consists of nodes (e.g., laptops, smart devices, sensors, VMs), living either on the *Cloud* or on the *Edge*, representing possible resources for our application. Regardless of the actual physical location of a resource, we are interested on its computational and storage capabilities. For an exhaustive description of the hardware characteristics of a node, standard approaches like the DMTF-CIM[1] could be adopted. Regardless of the specific model adopted, this information is required for understanding if a resource is suitable for hosting the execution of a tasks or for storing some data produced or consumed during the execution. The distinction between Cloud and Edge resources influences the level of detail with which a designer can describe a resource. A Cloud resource is managed by an external entity, i.e., the cloud provider, thus only a limited set of information

---

[1] http://www.dmtf.org/standards/cim.

is available for the resource. Conversely, for resources living on the Edge, we assume that the designer knows all the details about the resources and it has the ability to reconfigure or to adapt them in case modifications are required during the execution. Referring to our running example, we can assume that the designer has the possibility to change – both decreasing or increasing – the sampling rate for the sensors to a value that is optimal for the application. Conversely, the way in which the weather data are collected cannot be modified by the designer and we can also assume it is not feasible also for the Cloud provider hosting such data sources as they could be shared with other customers.

Data Source. The data source models the information needed to allow the application to read or write the relevant data. A data source is described in terms of the data content (i.e., how the data are structured), data access (i.e., how to reach data), and data utility (i.e., the relevance for the usage context). A data source is also defined by the *initialResource* where the source is initially deployed, as well as the *possibleResources* in which it could be moved as they satisfy the functional constraints required by the data source (e.g., the size, the DBMS or the file system). Inspired by the approach proposed in [7], data sources are distinguished between *Internal Sources* and *External Sources*. In the former case, we have data that are directly managed by the application designer. This category includes data produced by a task and consumed either by another task or by the final user (i.e., the $E_{i,j}$ data in the running example), data coming from resources managed by the designer (i.e., temperature sensors at the edge), or data sources on the Cloud where their management is under the responsibility of the designer (but the management of the resources on which the data source is deployed is up to the Cloud provider). In case of *External sources*, data exist independently of the application, but need to be accessed by it (e.g., the Weather data DS_W). In the proposed model, we assume that data content and data access are independent from the resource in which the data source is placed. This could be possible if a proper abstraction level is adopted to describe the intensional schema of the data, as also suggested in [2]. About the data access, using a proper naming scheme, like the one based on URI, makes the location transparency possible. Thus, when data source moves from a resource to another among the possible ones, this does not affect the behavior of tasks using these data. Data access also includes the definition of how to interact with the data source, distinguishing between stream or conventional methods. The focus of this paper concerns the *Data Utility* that measures to which extent a data source is relevant for a given usage. In the model, *Data Utility* (DU) is seen as a characterization of the association between a *Task* and a *Data Source* related to the input of the task. As discussed in Sect. 4, data utility of a data source extends the more traditional data quality concept. Differently from data content and access, DU could be affected by the resources used to host the data source. The DU model introduced in the next section makes this dependency explicit.

Task. A task represents a unit of work to be performed during the execution of the application. As suggested in [14], especially in data-intensive applications, there is a set of common tasks like data cleaning, data integration, data compression, data encryption, and the application can be seen as a composition of such tasks. In some cases, the algorithms behind those tasks are well known (e.g., clustering, regression), in some others the designer has to produce some scripts implementing the custom data processing. As for the data sources, there is a set of *possibleResources* on which the task can be deployed and one of them represents the *initialResource* where the task is initially deployed.

We assume that tasks are organized according to a data-flow process [5] which highlights how data are acquired, transformed, and returned. This flow is defined using *next* and *previous* attributes which capture the possible execution flows. The connection between the tasks and the data sources represents the input or the output of the tasks. In case of *output*, the task is connected to the storage nodes, as they represent data produced internally to the application. On the other side, the *input* of a task can be modeled both as a storage node or a source node as the input of a task can be data produced by a preceding task or made available by an external element.

4   Data Utility Model

Based on the data-intensive application model introduced in Sect. 3, now we go into the details of some of the concepts which constitute the elements for defining the Data Utility.

Starting from the available data sources, we indicate them with:

$$DS = \{ds_j\} = \{< S_j, ir_j, PR_j >\}$$

where S_j is the data source schema, ir_j is the resource on which ds_j is initially deployed, and PR_j is the set of possible resources on which it could be deployed. As the initial resource is, by definition, one the possible resource then $ir_j \in PR_j$.

Moving to the tasks that compose the data-intensive application, we assume that each task t_i is defined by:

$$t_i = < D_i, IN_i, OUT_i, P_i, N_i, ir_i, PR_i >$$

where *(i)* D_i is a description of the task in terms of type of operations performed (e.g., aggregation, filtering, clustering, association), *(ii)* IN_i and OUT_i are the sets of task inputs and outputs, *(iii)* P_i and N_i are the set of tasks that precede and follow the analyzed task in the data-flow process, *(iv)* ir_i refers to the initial resource on which the task is deployed, while PR_i the set of resources on which it can be potentially deployed. Similarly to what stated for the data sources, $ir_i \in PR_i$.

As described in Sect. 2, tasks may gather inputs (i) from a specific data source (i.e., *Task A*), (ii) from a previous task (i.e., *Task B*) and (iii) from a data source that should be selected from a set of candidate sources (i.e., *Task C*). Note that we consider cases (i) and (ii) as equivalent since we assume that the output of a task can be seen a data source. Furthermore, case (iii) is the situation in which the developer should be supported in the selection of the sources.

A task t_i may have several inputs. The k-th input is defined as:

$$IN_{ik} = < A_{ik}, CDS_{ik} >$$

where, A_{ik} is the set of the attributes of the data source required by the task (e.g., temperature, humidity), and $CDS_{ik} \subseteq DS$ the set of candidate data sources from which data have to be extracted, which can be both internal and external sources. If $|CDS_{ik}| = 1$, it means that the designer has specified the specific source to consider; otherwise the designer would like to be supported in the identification of the most suitable source in the specified set. In this last situation, the developer specifies a request R_{ik} over the input IN_{ik} in order to provide all the elements that can affect the source selection. Let us define the request as:

$$R_{ik} = < IN_{ik}, f_{ik}^*, NF_{ik}^* >$$

where, f_{ik} is an optional parameter to express functional requirements, while NF_{ik} is an optional parameter expressing a set of required non-functional properties. More precisely, f_{ik} is a predicate composed of atoms linked by traditional logical operators (i.e., AND, OR) that allows developers to specify restrictions over the allowed values in order to better drive the source selection (e.g., city= "Milan" AND Temp > 23). On the other hand, NF_{ik} contains requests related to DQ (Data Quality) or QoS (Quality of Service) aspects. The former focuses on the quality of the content provided by the source, while the latter regards performance issues such as availability and latency.

It is worth noting that the satisfaction of functional requirements only depends on the content of the data source, whereas the satisfaction of the QoS constraints depends on the resources on which the task or the data are deployed/stored. Therefore the suitability of a data source has to be specified by considering not only the data it contains but also the execution environment.

In this paper, for defining this suitability we introduce the Data Utility concept. *Data Utility (DU)* can be defined as *the relevance of data for the usage context*, where the context is defined in terms of the designer's goals and system characteristics. The designer's goals are captured by the definition of t_i which includes the input descriptions and the related requests in terms of both functional and non-functional requirements, while the system characteristics include the definition of the data sources DS. On these basis, Data Utility of a data source ds_j for a task t_i is defined as:

$$DU_{ixjy} = f(< t_i, r_x >, < ds_j, r_y >)$$

Since both tasks and data sources can be placed on different resources belonging to PR_i and PR_j, respectively, data utility depends on the task ($< t_i, r_x >$) and data sources ($< ds_j, r_y >$) placement, where $r_x \in PR_i$ and $r_y \in PR_j$.

We assume that the data sources are associated with a set of metadata that reveal the *Potential Data Utility* (PDU) that summarizes the capabilities of the data source and can be periodically evaluated independently of the context. The PDU is calculated looking at the data and the characteristics of the data source. It is derived from a *Data Quality* and a *Reputation* assessment. Generally speaking, as in [19] data quality can be defined as the fit for use for a data consumer and it implies a multi-dimensional analysis including dimensions like accuracy, completeness, timeliness [1]. In fact, errors, missing, or updated data affect the usage and potential benefits of data. The assessment of Data Quality dimensions may contribute to the understanding of the potential value of the data. The list of dimensions and the assessment metrics depend on the type of data contained in the source. For example, sensors data need the evaluation of additional attributes such as precision and data stability and algorithms for evaluating accuracy change along the type of data (i.e., strings vs. numeric values). Anyway, we can assume that each source is associated with a set of Data Quality dimensions and related values. Besides the content, also the history about the usage of the source should be considered. For this reason we define a Reputation index as the likelihood that a data source will satisfy the application requirements. For now, we compute the reputation by considering the frequency with which the source has been used and the respective success rate, and the scope of data (e.g., generic or specific, integrable with other sources, used with other sources). PDU provides an objective way to rank similar sources and can be useful for a pre-filtering of the sources. However, a data source has to be evaluated by considering the context that in our scenario is composed of the data-intensive application and the available resources. Given a request R_{ik} together with the characteristics of the task and the set of candidate data sources, the request can be enriched with additional data quality constraints derived by the type of task (e.g., data mining operations requires a high amount of data and completeness). The task type and request may also force the recalculation of the some data quality dimensions (i.e., if the request is limited to a subset of attributes of the source, the quality should be evaluated only on the considered data set).

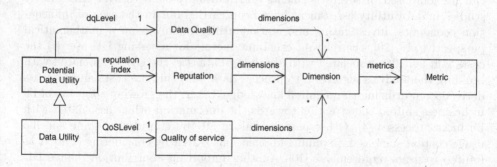

Fig. 3. Model of the utility components

QoS capabilities have to be evaluated by considering all the available options that the Fog environment offers. Thus, both tasks and data can be moved from edge to cloud and vice-versa, from edge to edge or through cloud resources and that the placement of a task or a data source on a specific resource has surely an impact on the QoS: in fact, the computational cost for obtaining data and the latency changes on the basis of the chosen location. Therefore, we calculate the QoS dimensions for each possible configuration defined in terms of task placement $< t_i, r_x >$ and data placement $< ds_j, r_y >$.

In summary, Data Utility can be assessed by considering three main aspects (Fig. 3): Data Quality, Reputation, and Quality of Service. Each of them is evaluated by means of dimensions, each one associated with different metrics (more than one assessment function might be available for a single dimension). Discarding the sources and configurations that do not satisfy functional and non-functional requirements, it is possible to associate with each source ds_j belonging to CDS_{ik} different Data Utility indicators (one for any admissible configuration), each one expressed as a set of three indices: Data Quality, Reputation and QoS.

5   Related Work

Managing data, meta-data and their storage and transformation has been addressed in several areas of research focusing on a number of separate though possibly interrelated aspects. Data utility has been defined in different ways in the literature. In statistics data utility is defined as "A summary term describing the value of a given data release as an analytical resource. This comprises the data's analytical completeness and its analytical validity" [9]. In business scenarios data utility is conceived as "business value attributed to data within specific usage contexts" [16] while in IT environments it has been described as "The relevance of a piece of information to the context it refers to and how much it differs from other similar pieces of information and contributes to reduce uncertainty" [11]. All these definitions agree on the dependency of DU on the context in which data are used. Therefore, the assessment of DU is a complex issues since context can be composed of several elements and it usually changes over time. Early studies in data utility assessment have been carried out in the area of information economics, investigating information utility mainly from a mathematical perspective. In [20], the relevant economic factors for assessing DU are: (i) the costs and benefits associated with obtaining data, (ii) the costs associated with building the analysis algorithm to process data, and (iii) the costs and benefits derived from utilizing the acquired knowledge. Later, the growing adoption of IT in business shifted the attention towards the information utility associated with business processes [4]. Other papers analyze DU by considering some specific usage context such as data mining applications [12] or by considering context as limited to users requirements [10]. Another important contribution relates DU to information quality dimensions, e.g. accuracy and completeness [13]. Information quality requirements for obtaining valuable results from processes have been discussed in [6]. In [18], data utility is discussed in mobile clouds with the focus on optimizing the energy efficiency of mobile devices. Energy efficiency was also the focus of [8] where the interrelations between data value evaluation and adaptation strategies have been discussed, with a focus on run-time adaptation rather than on the design of applications.

We propose a more comprehensive definition of Data Utility that includes both the content of the data sources, the application using them and the execution environment, considering all the data and computation movement actions that fog computing enables. Note that the concept of data movement has been discussed in [3] as a basis for providing operations for improving quality of data and service. However, while that paper focuses on possible operations and strategies, in the current paper we focus on a comprehensive evaluation of data utility.

6 Concluding Remarks

In this paper we have introduced a conceptual model to define the Data Utility for data-intensive applications in a Fog environment. The proposed model takes into account the relationship between the tasks composing the application and the data sources that can be used by such tasks to perform the required computation. As the location of both tasks and data sources can change, the influence of data and computation movement is considered in the Data Utility Model. The evaluation of the Data Utility and the definition of a Global Utility Model for the whole application is under investigation to consider the influences among tasks, as well as the constraints over the deployment.

Acknowledgments. DITAS project is funded by the European Union Horizon 2020 research and innovation programme under grant agreement RIA 731945.

References

1. Batini, C., Scannapieco, M.: Data and Information Quality - Dimensions, Principles and Techniques. Data-Centric Systems and Applications. Springer, Cham (2016). https://doi.org/10.1007/978-3-319-24106-7
2. Cleve, A., Brogneaux, A.-F., Hainaut, J.-L.: A conceptual approach to database applications evolution. In: Parsons, J., Saeki, M., Shoval, P., Woo, C., Wand, Y. (eds.) ER 2010. LNCS, vol. 6412, pp. 132–145. Springer, Heidelberg (2010). https://doi.org/10.1007/978-3-642-16373-9_10
3. D'Andria, F., Field, D., Kopaneli, A., Kousiouris, G., Garcia-Perez, D., Pernici, B., Plebani, P.: Data movement in the Internet of Things domain. In: Dustdar, S., Leymann, F., Villari, M. (eds.) ESOCC 2015. LNCS, vol. 9306, pp. 243–252. Springer, Cham (2015). https://doi.org/10.1007/978-3-319-24072-5_17
4. Even, A., Shankaranarayanan, G., Berger, P.D.: Inequality in the utility of customer data: implications for data management and usage. J. Database Mark. Custom. Strat. Manag. **17**(1), 19–35 (2010)
5. Garijo, D., Alper, P., Belhajjame, K., Corcho, Ó., Gil, Y., Goble, C.A.: Common motifs in scientific workflows: an empirical analysis. Future Generat. Comp. Syst. **36**, 338–351 (2014)
6. Gharib, M., Giorgini, P., Mylopoulos, J.: Analysis of information quality requirements in business processes, revisited. Requir. Eng. 1–23 (2016)
7. Gómez, A., Merseguer, J., Di Nitto, E., Tamburri, D.A.: Towards a UML profile for data intensive applications. In: Proceedings of the International Workshop on Quality-Aware DevOps, Saarbrücken, Germany, pp. 18–23 (2016)

8. Ho, T.T.N., Pernici, B.: A data-value-driven adaptation framework for energy efficiency for data intensive applications in clouds. In: 2015 IEEE Conference on Technologies for Sustainability (SusTech), pp. 47–52. IEEE (2015)

9. Hundepool, A., Domingo-Ferrer, J., Franconi, L., Giessing, S., Nordholt, E.S., Spicer, K., de Wolf, P.P.: Statistical Disclosure Control. Wiley, New York (2012)

10. Ives, B., Olson, M.H., Baroudi, J.J.: The measurement of user information satisfaction. Commun. ACM **26**(10), 785–793 (1983)

11. Kock, N.F.: Encyclopedia of E-collaboration. Information Science Reference - Imprint of: IGI Publishing, Hershey (2007)

12. Lin, Y.C., Wu, C.W., Tseng, V.S.: Mining high utility itemsets in big data. In: Cao, T., Lim, E.P., Zhou, Z.H., Ho, T.B., Cheung, D., Motoda, H. (eds.) PAKDD 2015. LNCS, vol. 9078, pp. 649–661. Springer, Cham (2015). https://doi.org/10.1007/978-3-319-18032-8_51

13. Moody, D., Walsh, P.: Measuring the value of information: an asset valuation approach. In: European Conference on Information Systems (1999)

14. Nalchigar, S., Yu, E., Ramani, R.: A conceptual modeling framework for business analytics. In: Comyn-Wattiau, I., Tanaka, K., Song, I.Y., Yamamoto, S., Saeki, M. (eds.) ER 2016. LNCS, vol. 9974, pp. 35–49. Springer, Cham (2016). https://doi.org/10.1007/978-3-319-46397-1_3

15. OpenFog Consortium Architecture Working Group: OpenFog Architecture Overview, February 2016. http://www.openfogconsortium.org/ra

16. Syed, M.R., Syed, S.N.: Handbook of Research on Modern Systems Analysis and Design Technologies and Applications. Information Science Reference - Imprint of: IGI Publishing, Hershey (2008)

17. Turner, V., Reinsel, D., Gatz, J.F., Minton, S.: The digital universe of opportunities. IDC White Paper, April 2014

18. Wang, J., Zhu, X., Bao, W., Liu, L.: A utility-aware approach to redundant data upload in cooperative mobile cloud. In: 9th IEEE International Conference on Cloud Computing, CLOUD 2016, San Francisco, CA, USA, pp. 384–391 (2016)

19. Wang, R.Y., Strong, D.M.: Beyond accuracy: what data quality means to data consumers. J. Manage. Inf. Syst. **12**(4), 5–33 (1996)

20. Weiss, G.M., Zadrozny, B., Saar-Tsechansky, M.: Guest editorial: special issue on utility-based data mining. Data Min. Knowl. Discov. **17**(2), 129–135 (2008)

Assessing the Positional Planimetric Accuracy of DBpedia Georeferenced Resources

Abdelfettah Feliachi[1]([⊠]), Nathalie Abadie[1], and Fayçal Hamdi[2]

[1] LaSTIG-COGIT, Université Paris-Est, IGN/SRIG, Saint-Mandé, France
`abdelfettah.feliachi@ign.fr`
[2] CEDRIC - Conservatoire National des Arts et Métiers, Paris, France

Abstract. Assessing the quality of the main linked data sources on the Web like DBpedia or Yago is an important research topic. The existing approaches for quality assessment mostly focus on determining whether data sources are compliant with Web of data best practices or on their completeness, semantic accuracy, consistency, relevancy or trustworthiness. In this article, we aim at assessing the accuracy of a particular type of information often associated with Web of data resources: direct spatial references. We present the approaches currently used for assessing the planimetric accuracy of geographic databases. We explain why they cannot be directly applied to the resources of the Web of data. Eventually, we propose an approach for assessing the planimetric accuracy of DBpedia resources, adapted to the open nature of this knowledge base.

1 Context and Objectives

In the Web of data, many resources are associated with some location on Earth, either directly (using coordinates or geometric primitives) or indirectly (with an address or a place name). Direct spatial references can be used for spatial data analysis, cartographic visualization or to georeference other resources. Like any other properties, spatial references can also be used to evaluate the similarity of the resources they describe for data linking purposes. Two resources of similar types described by similar spatial references may thus be considered as representing the same real world entity and therefore be linked to each other.

Each of the possible use cases involving the spatial references associated with the data requires taking into account their absolute positional planimetric accuracy, that is to say the difference between the locations provided by the spatial references and the locations considered as true [9]. [1] defines minimal positional planimetric accuracy values that Web gazetteer data should respect to provide accurate spatial references for resources that reuse their coordinates. However, information about the quality of these spatial references, often provided in the metadata of geographic datasets is, to the best of our knowledge, almost never available for resources published in the Web of data. Their evaluation seems little considered since it does not appear in the main state-of-the-art works dealing with the Web of data quality issues, such as [14].

© Springer International Publishing AG 2017
S. de Cesare and U. Frank (Eds.): ER 2017 Workshops, LNCS 10651, pp. 227–237, 2017.
https://doi.org/10.1007/978-3-319-70625-2_21

This issue of assessing the quality, reliability and credibility of the Web of data resources is at the origin of the upper layers of the semantic Web "layer cake". Associating credibility ("Trust" layer) to data is based on provenance information and the facts inferred from it ("Proof" and "Logic" layers). Hence, W3C has published PROV [7], a set of recommendations for exchanging inter-operable provenance data on the Web. They include a conceptual data model, the associated OWL2 ontology, a serialization text language, data integrity constraints, and so on.

In this article, we propose an approach to evaluate the positional planimetric accuracy of spatial references associated with the Web of data resources, adapted to the open nature of the Web data sources. We perform tests on resources extracted from the French DBpedia[1] that describe the monuments of Paris.

2   Existing Approaches for Evaluating the Absolute Positional Planimetric Accuracy of Georeferenced Data

2.1   Approaches for Geographic Databases

Direct spatial references[2] are used to provide a quantitative description of the characteristics of the real-world geographic entities, such as their location, shape, size or orientation. The representation of these characteristics through geometries depends on two main factors: the level of detail expected for the database, that is to say the level of geometric and semantic abstraction used in this database to represent real-world geographic entities [11], and the limitations dues to the resolution of the raw data sources used for geometry capture. Both may lead to simplified representations of the real world entities and thus to notable differences from one database to another [5].

Rules, describing how geographic entities should be represented by geometries with boundaries captured along some given characteristic element of their shape, are provided to data-entry operators in order to guarantee a good homogeneity of the captured geometries. Besides, the absolute positional planimetric accuracy of the geometries depends strongly on the raw data used for their capture. In the case of geographic data provided by traditional data producers (e.g. national mapping agencies), the whole geometry acquisition process is designed to obtain data with a predefined absolute positional planimetric accuracy.

ISO 19157 standard on geographic data quality distinguishes two types of quality evaluation methods. The indirect methods are based on knowledge about data provided either by external sources or by the experience gained about data possibilities and limitations. The external knowledge sources may be qualitative metadata or genealogic information. This is closely related to the motivations of the PROV [7] recommendation: providing knowledge about the data lifecycle to

---

[1] http://fr.dbpedia.org/. Data extracted in December 2013 and containing 625 resources.

[2] For the sake of brevity, we will use the term "geometries" instead of "direct spatial references" in the remainder of this article.

assess the quality of a data source. Direct evaluation methods are based on the inspection of the data. Data may be analyzed on their own (absolute method) or they may be compared with other data sources (relative method). Relative methods require a high quality reference data source. These methods can be applied to the entire dataset or to a representative sample. The catalog of standardized quality measures of the ISO 19157 standard offers many numerical methods to evaluate the positional planimetric accuracy of the geometries. However, these methods are only applicable to data sources with a quite homogeneous data capture process. Otherwise, the standard recommends categorizing data by cause of heterogeneity and applying the chosen method independently to each subset.

2.2   Approaches for Linked Georeferenced Data

Unlike in geographic databases, geometries associated with Web of data resources are not the main piece of information in their description. In recent years, many vocabularies have been proposed to represent geographic features geometries on the Web [2] and standardization work is currently under way [12]. Geometries may come from geographic databases converted to RDF and published by national mapping agencies[3]. But most data sources on the Web include geometries of various provenances: Geonames (http://www.geonames.org/) gathers several traditional geographic datasets with crowdsourced data and the large georeferenced data sources DBpedia, Yago and LinkedGeoData are derived from the crowdsourcing projects Wikipedia and OpenStreetMap. When the data come from various sources, geometry acquisition processes are less controlled or less known. It may then be difficult to assess the positional planimetric accuracy of geometries, which can vary significantly from one geometry to another.

Works on linked data quality focus on their compliance with good practices in the Web of data. The survey on the quality assessment measures proposed by [14] gives priority to measures on dereferencing, licensing and interconnecting issues. Then come measures about the intrinsic quality of data, their fitness for user's needs, and their representation. However, none of the presented measures addresses the quality of the spatial references associated to resources. Nevertheless, the approach for the detection of aberrant numerical values in DBpedia proposed by [13] allows the identification of outliers in coordinates or altitudes values. In addition, [1] evaluates the positional planimetric accuracy of GeoNames coordinates with respect to their number of decimals.

Assessing the quality of the location information produced by crowdsourcing projects is a key issue addressed in many studies. [8] proposes an approach to evaluate the positional planimetric accuracy of geotags associated with FlickR images of remarkable buildings. The positional planimetric accuracy of the geotags is evaluated by calculating the average distance between their coordinates and those provided by the Wikipedia article describing the photographed buildings. The choice of Wikipedia as a reference data source for buildings location

---

[3] This is the case for data published by the Ordnance Survey http://data.ordnancesurvey.co.uk or IGN Spain http://geo.linkeddata.es/.

is, unfortunately, not discussed. Finally, OpenStreetMap (OSM) is probably the volunteered geographic data source whose quality has been the most extensively studied [3,6]. [3] provides, for the OSM data, a set of indirect quality measures based on the available genealogy metadata. [6] evaluates the quality of OSM data on French territory using a set of standard direct measures. The positional planimetric accuracy of three OSM data samples (respectively with point, polyline and polygon geometries) is estimated by computing the mean distance between each geographic feature of these samples and their counterparts retrieved from a reference dataset produced by the French national mapping agency.

3   Genealogy of DBpedia Resources Geometries

DBpedia resources describing real-world geographic entities are georeferenced using points extracted from Wikipedia. Therefore, their positional planimetric accuracy depends directly on the coordinates provided by the Wikipedia contributors. Nearly 15% of Wikipedia articles are georeferenced and 16.25% derive their coordinates from Wikidata[4].

The "WikiProject Geographical Coordinates"[5] aims to improve the quality of the Wikipedia articles coordinates by providing recommendations for their capture. First, they advocate the use of trusted sources, such as the geoportals of national mapping agencies, to find reliable coordinates. In addition, they indicate which characteristic shape element should be localized, depending on the type of geographic entity described by each article: the center of the inhabited area for municipalities or the main entrance for buildings. Recommendations also provide rules to round the coordinate values in order to have a coordinates precision consistent with the size of the geographic entities. For example, the coordinates[6] of a geographic entity located in France with a length between 50 and 100 m should ideally have 4 decimals when they are expressed in decimal degrees.

These recommendations explain how to add these coordinates in the body of a Wikipedia article or in its infobox using the predefined template "Template:Coord"[7]. In addition to coordinates, it includes metadata. Most of them, such as "type", "dim" or "scale", are intended to define the most appropriate map scale to visualize the geographic entity. The metadata "source" indicates the provenance of the entered coordinates: coordinates obtained from the *Geographic Names Information System* should therefore mention "source:GNIS". The geographic coordinate extractor used by the French DBpedia[8] searches for this template in Wikipedia articles and infoboxes and only keeps the coordinates to produce triplets based on geo[9]:long, geo:lat and georss:point properties.

---

[4] Source: https://fr.wikipedia.org/wiki/Projet:Géolocalisation.

[5] https://en.wikipedia.org/wiki/Wikipedia:WikiProject_Geographical_coordinates.

[6] Expressed in the WGS84 coordinates reference system.

[7] https://en.wikipedia.org/wiki/Template:Coord.

[8] http://fr.dbpedia.org/doc/listeExtracteurs.html.

[9] http://www.w3.org/2003/01/geo/wgs84_pos#.

There is no guarantee that all Wikipedia contributors are aware of these recommendations and that they apply them correctly. Similarly, there is no evidence that they rely on data from national mapping agencies to determine the coordinates associated with articles or that they follow the recommendations about coordinates precision. Finally, nothing forces them to fill out the "Template: Coord" metadata, and even if they do, the DBpedia coordinates extractor does not keep them. Direct quality methods recommended by ISO 19157 were designed to evaluate the overall positional planimetric accuracy of homogeneous datasets. These methods are not directly applicable here. In addition, with no reliable genealogy metadata, indirect assessment methods cannot be used.

4   Direct Evaluation of the Absolute Positional Planimetric Accuracy of DBpedia Resources

In order to overcome the lack of genealogy metadata and to provide a single positional planimetric accuracy estimation for each resource, we propose to adapt the direct methods designed to assess the absolute positional planimetric accuracy of traditional geographic data to the specific case of Web of data resources. These require to be treated on a case-by-case basis.

4.1   The Proposed Approach

We propose a two-step approach. The first step aims to find, for each resource to be evaluated, which characteristic element of its shape was pointed out to define its coordinates. Then, the distance between each DBpedia point and its supposed counterpart within a reference geographic dataset is computed.

Using a point to locate a geographic entity provides a highly simplified geometric representation of that entity. Moreover, this requires deciding what characteristic element of its shape should be pointed out preferably. Therefore the evaluation of the positional accuracy of such a point cannot be done without taking into account this representation choice: its coordinates must be compared with those of a point captured in the same way, but with a higher accuracy.

In the case of historical monuments, the recommendations made by the project "WikiProject Geographical Coordinates" seem to indicate to localize the entrance of each monument, preferably by picking its coordinates on the relevant national mapping agency geoportal. However, if we plot the coordinates of Paris DBpedia monuments on an IGN orthophotographic base map, we observe significant shifts with respect to the expected localization (see Fig. 1). It seems therefore impossible to rely on the recommendations of the project "WikiProject Geographical Coordinates" to determine which point of the shape of historical monuments is represented by DBpedia coordinates.

We therefore propose to formulate hypotheses on the choices made by contributors when entering the coordinates of the monuments. Then, we compare monuments coordinates with points corresponding to each category of representation choice, selected from a reference geographic dataset. This provides us with

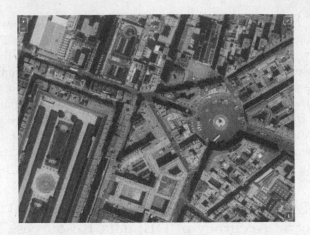

Fig. 1. DBpedia historical monuments, in yellow, on IGN orthophotographic base map. (Color figure online)

indicators for classifying each monument coordinates by category of representation choice. This step can be carried out with common spatial analysis tools and a supervised classification method.

The second step of our approach consists in comparing the evaluated coordinates with those of the points identified as corresponding to the representation choice made by the contributors and selected within a reference geographic dataset having a better and well documented positional planimetric accuracy.

4.2   Implementation

When the points used for locating DBpedia historic monuments are plotted on an IGN orthophotographic base map, three types of representation choices can be identified: close to the center of the building considered as a historical monument, close to its facade and finally near the road centerline in front of its facade. The two first representation choices correspond respectively to two types of recommendations from the "WikiProject Geographical Coordinates": for geographical entities with a broad spatial extent coordinates should be captured at their center, and for buildings at their main entrance. The third representation choice is a common practice for capturing addresses.

Relevant spatial indicators must then be defined to decide to which category of representation choice belongs each point [10]. We used two IGN geographic datasets, chosen for their consistency with the "WikiProject Geographical Coordinates" guidelines: the buildings from the BD PARCELLAIRE®[10] and the roads from the BD TOPO®[11]. Using the PostgreSQL/PostGIS database management system, we calculated three indicators based on the distance between

---

[10] Cadastral database produced by the IGN.

[11] Database describing the topography of the French territory and its infrastructures produced by the IGN.

the DBpedia points and the reference geometries: the distances to the barycenter and to the facade of the nearest building and the distance to the nearest road segment. Since the sizes of the buildings and the roads vary considerably according to the districts of Paris, we normalized these values (see Fig. 2).

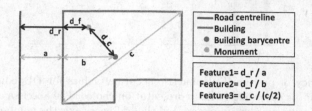

Fig. 2. Learning features for geometry capture rules.

We then manually prepared a learning sample of about 30 monuments of each type. Finally, we used Weka[12] as it implements the most commonly used supervised classification algorithms. We applied several of the algorithms available in Weka to our data and we manually checked the results.

Finally, we computed the absolute positional planimetric accuracy of each DBpedia point by adding two values: its distance to the point in the geographic reference dataset identified as the representation choice made by its contributor - the building facade, the building barycenter or the road centerline - and the absolute positional planimetric accuracy of the reference geographic dataset.

4.3   Results and Discussion

The Table 1 presents the results of the four classification algorithms tested in order to assign to each DBpedia point a category of representation choice (see [4] for details about the classifiers). These results are compared with a manual classification. The four tested algorithms provide good results, which tends to validate our choice of indicators.

Table 1. Learning results for some applied classification algorithms.

Method	Precision	Rappel	F-measure
Bayes network	91,6%	91,3%	91,3%
JRIP	96,3%	96,3%	96,3%
Decision table	96,4%	96,3%	96,2%
Random forest	98,8%	98,8%	98,7%

---

[12] http://www.cs.waikato.ac.nz/ml/weka/.

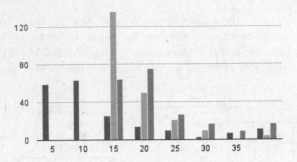

Fig. 3. Frequency of positional planimetric accuracy values for DBpedia's historical monuments in Paris according to the representation choice. The abscissa axis indicates the maximum accuracy values in meters and the ordinates axis the number of DBpedia resources. The yellow, mauve, and blue sticks represent the resources captured respectively at the road centerline, at the facade and at the building barycenter. (Color figure online)

Figure 3 shows the distribution of the obtained positional planimetric accuracy values. The values of the estimated positional uncertainties are mostly low, with a strong predominance of values between 15 and 25 m. The resources captured at the facade or at the center of buildings have planimetric accuracy values greater than 10 m due to the relatively large average planimetric accuracy value given by the BD PARCELLAIRE® metadata[13]. Those captured at the road centerline have relatively low values. This is probably due to spatial indicators used for the classification that require relatively small distances to the road centerline to assign a resource to this class, as well as the very low absolute planimetric accuracy values of the BD TOPO® for Paris road segments.

Figure 4 shows the results, represented by circles centered on the DBpedia monuments and of radius equal to their respective accuracy values. The monuments located by their barycenter have greater positional uncertainties than the others, which confirms the distribution of the Fig. 3.

The results strongly depend on the initial assumptions underlying the overall evaluation. The categories of representation choices are defined by comparing the coordinates of DBpedia resources with geographic data recommended by the "WikiProject Geographical Coordinates" as reference sources. We thus assume that IGN databases have actually been used as coordinates sources by all Wikipedia contributors. Additionally, we assume that the coordinates of the DBpedia resources describing historical monuments are accurate enough so that their closest building in IGN databases can be considered as representing the same monument. This also assumes that DBpedia coordinates possess 4 decimals (or even 5 for the smallest buildings) as recommended by the "WikiProject Geographical Coordinates". From the 625 historical monuments analyzed, 606 have coordinates with at least 4 decimal places and 500 coordinates with at least 5 decimal places. This tends to confirm that the coordinates capture

---

[13] http://professionnels.ign.fr/sites/default/files/DC_BDPARCELLAIRE_1-2.pdf.

Fig. 4. Positional planimetric accuracy values of DBpedia monuments.

recommendations are rather respected on this point and that contributors were motivated to provide accurate spatial information. On the other hand, the distribution of resources in the three categories of representation choices tends to show that the capture recommendations about the characteristic element of the shape to be represented are not followed. In fact, resources are almost equally distributed between the three categories of choice in the manual classification: 33.4% for the buildings barycenter, 35.4% for their facade and 31.2% for the road centerline. The first step of our approach, which aims to find for each resource what choice of representation has been made, is therefore essential.

Our approach is particularly suited to data sources that represent geographic entities, distinguishable as individual topographic objects, by points captured at the level of a characteristic element of their shape, a priori unknown and potentially different for each resource. It seems to be applicable for linear geographical entities as Wikipedia capture recommendations[14] also encourage to represent them by points captured at the level of well defined shape characteristic elements. On the other hand, it is less applicable for geographic entities perceived by aggregation of individual objects, such as urban areas.

In order to implement our approach for each type of georeferenced resource, the possible categories of representation choices must be identified, spatial indicators must be defined and computed and learning samples must be created for each of category of representation choices to be considered. A first step towards its generalization to all the categories of DBPedia georeferenced resources could be to adapt it to different samples chosen for the variety of their types of representation choices and the required spatial indicators.

---

[14] https://en.wikipedia.org/wiki/Wikipedia:WikiProject_Geographical_coordinates/Linear.

5   Conclusion and Perspectives

The evaluation of the positional planimetric accuracy of the geometries associated with the DBpedia resources made us study their genealogy and compare the contributors practices with the capture recommendations formulated by the Wikipedia project. In the case of historical monuments in Paris, it appears that the recommendations on the number of decimals of the coordinates seem to be respected. Similarly, the predominantly low positional planimetric accuracy values suggest that contributors use accurate data sources. On the other hand, the capture recommendations concerning the choice of the shape characteristic element of the entity to be localized seem much less followed.

Providing georeferenced resources with genealogy information on the coordinates source and the choices of representation would make indirect evaluations of their positional planimetric accuracy possible or simplify the direct evaluation of this accuracy. In addition, such metadata would be useful to implement spatial analysis applications aware of the potential and limitations of the geometries associated with Web resources. For these purposes, extensions of vocabularies such as PROV-O or DQV[15] using the ISO 19157 [9] standard could be considered.

References

1. Ahlers, D.: Assessment of the accuracy of geonames gazetteer data. In: Proceedings of the 7th Workshop on Geographic Information Retrieval, pp. 74–81. ACM (2013)
2. Atemezing, G.A., Troncy, R.: Comparing vocabularies for representing geographical features and their geometry. In: Terra Cognita 2012 Workshop, vol. 3 (2012)
3. Barron, C., Neis, P., Zipf, A.: A comprehensive framework for intrinsic openstreetmap quality analysis. Trans. GIS **18**(6), 877–895 (2014)
4. George-Nektarios, T.: Weka classifiers summary. Athens University of Economics and Bussiness Intracom-Telecom, Athens (2013)
5. Girres, J.F.: Modèle d'estimation de l'imprécision des mesures géométriques de données géographiques. Application aux mesures de longueur et de surface. Thèse de doctorat, spécialité sciences et technologies de l'information géographique, Université Paris-Est (2012)
6. Girres, J.F., Touya, G.: Quality assessment of the french openstreetmap dataset. Trans. GIS **14**(4), 435–459 (2010)
7. Groth, P., Moreau, L.: Prov-overview, an overview of the prov family of documents. W3c working group note 30 April 2013, W3C (2013). https://www.w3.org/
8. Hauff, C.: A study on the accuracy of flickr's geotag data. In: Proceedings of the 36th International ACM SIGIR Conference on Research and Development in Information Retrieval, pp. 1037–1040. ACM (2013)
9. ISO: 19157: Geographic information - data quality. International standard, International Organization for Standardization (2013). http://www.iso.org
10. Mohri, M., Rostamizadeh, A., Talwalkar, A.: Foundations of Machine Learning. MIT Press, Cambridge (2012)

---

[15] https://www.w3.org/TR/vocab-dqv/.

11. Sarjakoski, L.: Conceptual models of generalisation and multiple representation. Generalisation of geographic information: cartographic modelling and applications. Elsevier, Amsterdam, The Netherlands (2007)
12. Tandy, J., Barnaghi, P., van den Brink, L.: Spatial data on the web best practices. W3c working group note 25 october 2016, W3C and OGC (2016)
13. Wienand, D., Paulheim, H.: Detecting incorrect numerical data in DBpedia. In: Presutti, V., d'Amato, C., Gandon, F., d'Aquin, M., Staab, S., Tordai, A. (eds.) ESWC 2014. LNCS, vol. 8465, pp. 504–518. Springer, Cham (2014). doi:10.1007/978-3-319-07443-6_34
14. Zaveri, A., Rula, A., Maurino, A., Pietrobon, R., Lehmann, J., Auer, S.: Quality assessment for linked data: a survey. Sem. Web **7**(1), 63–93 (2016)

Assessing the Completeness Evolution of DBpedia: A Case Study

Subhi Issa, Pierre-Henri Paris, and Fayçal Hamdi[✉]

CEDRIC - Conservatoire National des Arts et Métiers,
292 rue saint martin, Paris, France
Faycal.Hamdi@cnam.fr

Abstract. RDF web datasets, thanks to their semantic richness, variety and fine granularity, are increasingly adopted by both researchers' and business communities. However, as anyone can publish data, this leads to sparse and heterogeneous data descriptions with undeniably an impact on quality. Consequently, there is an increasing effort dedicated to Web data quality improvement. We are interested in data quality and precisely in completeness quality evolution over time. The paper presents a set of experiments aiming to analyze the evolution of completeness quality values over several versions of DBpedia.

1 Introduction

While earlier datasets were relatively homogeneous and reasonably small, data on the Internet age, is more like a huge patchwork collected from various and numerous sources [10]. This data, even rich in content, is often incomplete and lacks metadata leading to unreliable analyses.

In this paper, we are interested in how the completeness of a dataset evolve. We believe that understanding this evolution could help define more suitable strategies for data sources integration, enrichment and maintenance. According to [8], the data completeness is about to know to which extent a dataset contains all of the necessary objects for a given task. As a consequence, a dataset may fit for a usage but not for another, and it is almost impossible to have an absolute measure of its completeness. By combining several types of measures about completeness of a given dataset, one can approach its real completeness value. In this paper, we do not aim at measuring an absolute completeness but rather measure a facet or an aspect of it. Our intuition is the following: things sharing the same class are described with some common properties. For a given class (in the sense of RDF type[1], or a DBpedia category in our precise case) there are properties that every instance of this class *should* have.

Example 1. Each person has a name, a birth place, etc. So, this should be reflected in the data, all the person instances *should* have all those properties. But not all people have a death place (at least for the moment), so this property may not be found in all the person instances.

---

[1] https://www.w3.org/TR/rdf-schema/.

© Springer International Publishing AG 2017
S. de Cesare and U. Frank (Eds.): ER 2017 Workshops, LNCS 10651, pp. 238–247, 2017.
https://doi.org/10.1007/978-3-319-70625-2_22

Hence, if a property p is sufficiently used among a set of instances of the same class T, we postulate that this property p is kind of mandatory (or may exist in the reality even if not present in the dataset) for all instances of this class T. Therefore, if p is missing for an instance of T, we may reasonably argue that this decreases the completeness of the dataset. Some data mining approaches may help to find those important set of properties.

We conducted an exploratory analysis of data completeness along several classes from DBpedia. We first defined a completeness assessment method that we applied in three versions of DBpedia. The remainder of this paper is organized as follows: Sect. 2 summarizes a related literature on the subject, then Sect. 3 exhibit a motivating example, while Sect. 4 details the completeness calculation approach. Section 5 presents and analyzes a set of experiments. Finally, Sect. 6 draws conclusions and future research directions.

2   Related Works

Information quality attracted many research works for two decades. It is an interesting theoretical as well as practical domain and several projects proposed methodologies to help deal with quality assurance in traditional business information systems [1]. From the Web of Data perspective, the problem represents a new issue. Researchers pointed out the fact that making data accessible is not sufficient, especially when users are companies and governmental agencies, and when the target usage is business, research or countries' security. The credibility of data sources is thus bound to the quality of the content.

Several proposals could be classified using 4 categories from the TDQM[2] community [11] namely: *Intrinsic* (accuracy, reputation, believability and provenance), *Representational* (understandability, consistency and conciseness), *Accessibility* (accessibility and security), and *Contextual* (amount of data, relevance, completeness and timeliness). Intrinsic quality relies on internal characteristics of the data during evaluation. Most of the proposal concentrate on provenance data elicitation [5]. From an external point of view, representational quality is concerned with factors influencing users interpretations and practices such as understandability or data conciseness [12]. Concerning accessibility, a very important criterion in the context of web data, we could cite [7] where authors discussed common RDF publishers' errors that have a direct impact on data accessibility. Finally, contextual quality means that data could not be said "good" or "poor" without considering the context in which it is produced or used. The relevance dimension is considered from a ranking point of view in [2] and in the context of heterogeneous web data search in [6].

3   Motivating Example

We illustrate in this section the main idea behind our approach through an example that shows the issues and the difficulties encountered in the calculation of a dataset completeness. Let consider the set of scientists described in the

---

[2] Total Data Quality Management.

well-known open linked dataset, DBpedia. We would like to know, when users querying this dataset about a scientist (or a subset of scientists), if the information provided for this scientist are complete (well described) or not. Taking for instance a subset of 100 scientists. The pseudo-code 1 returns, for each scientist, the couples ⟨*property, value*⟩.

Algorithm 1. Scientists Descriptions

String *Query*1 = "`SELECT ?subject where{`
                  `?subject rdf:type dbo:Scientist`
                  `} LIMIT 100`"
Result S = ExecQuery(*Query*1)
for each *subject* ∈ S **do**
    String *Query*2 = "`SELECT ?property ?value where{`
                      `subject ?property ?value}`"
    Result R = ExecQuery(*Query*2)
    Descriptions.put(*subject*, < *property, value* >)
return *Descriptions*

To evaluate the completeness of this subset, a first intuition could consist of comparing the properties used in the description of each scientist with a reference scientist schema (ontology). For example, in DBpedia, the class *Scientist*[3] has a list of 4 properties (e.g. *doctoralAdvisor*), but these properties are not the only ones used in the description of a scientist (e.g. the *birthdate* property is not present in this list). Indeed, the class *Scientist* has a super class called *Person*. So, the description of a scientist may also take into account the properties of this class. Therefore, to obtain an exhaustive list of the whole properties used in the description of a scientist, we have to calculate the union of the set of properties of the class *Scientist* and all its ancestors. For our example, the reference scientist schema that we called *Scientist_Schema* could be calculated as follows:

$$Scientist\_Schema = \{Properties\,on\,Scientist\} \cup$$
$$\{Properties\,on\,Person\} \cup \{Properties\,on\,Agent\} \cup$$
$$\{Properties\,on\,Thing\}$$

such that: $Scientist \sqsubseteq Person \sqsubseteq Agent \sqsubseteq Thing$

Thus, the completeness of a scientist description (e.g. *Albert_Einstein*) will be the proportion of properties used in the description of this scientist to the total number of properties in *Scientist_Schema*. In the case of DBpedia, with a simple SPARQL query[4], we can obtain the size of *Scientist_Schema*, which

---

[3] http://mappings.dbpedia.org/server/ontology/classes/.
[4] Performed on: http://dbpedia.org/sparql.

is equal to 664 (A-Box properties). So, the completeness of the description of *Albert_Einstein* could be calculated as follows:

$$Comp(Albert\_Einstein) = \frac{|Properties\ on\ Albert\_Einstein|}{|Scientist\_Schema|}$$
$$= \frac{21}{664} = 4,21\%$$

However, for example, the property *weapon* is in *Scientist_Schema*, but is not relevant for the *Albert_Einstein* instance.

We can finally conclude that, the completeness as calculated here, does not provide us with the relevant value regarding the real representation of scientists in the DBpedia dataset. Hence, to overcome this issue, we may have to explore those instances. We want to find which properties are used more often than others to describe instances of a given type. Based on data mining, the approach that we propose in this paper, deals with this issue by extracting, from a set of instances (of the same class), the set of the most representative properties and calculates completeness in respect to this set.

4   Completeness Calculation: A Mining-Based Approach

To assess the completeness of the different version of a LOD dataset (as DBpedia), we propose an approach that calculates the completeness of an input dataset by posing the problem as an itemset mining problem. In fact, the completeness at the data level assesses missing values [9]. This vision requires a schema (e.g. a set of properties) that needs to be inferred from the data source. However, it is not relevant to consider, for a subset of resources, the schema as the union of all properties used in their description as seen in Sect. 3. Indeed, this vision neglects the fact that missing values could express inapplicability. Our mining-based approach includes two steps:

1. **Properties mining**: Given a dataset $\mathcal{D}$, we first represent the properties, used for the description of the $\mathcal{D}$ instances, as a transaction vector. We then apply the well-known FP-growth algorithm [4] for mining frequent itemsets (we chose FP-growth for efficiency reasons. Any other itemset mining algorithm could, obviously, be used). Only a subset of these frequent itemsets, called "Maximal" [3], is captured. This choice is motivated by the fact that, on the one hand, we are interested in important properties for a given class that should appear often and, on the other hand, the number of frequent patterns could be exponential when the transaction vector is very large (see Sect. 4.1 for details).

2. **Completeness calculation**: Once the set of maximal frequent itemsets $\mathcal{MFP}$ is generated, we use the apparition frequency of items (properties) in $\mathcal{MFP}$, to give each of them a weight that reflects how important the set of properties is considered for the description of instances. Weights are then

exploited to calculate the completeness of each transaction (regarding the presence or absence of properties) and, hence, the completeness of the whole dataset.

In the following we give a detailed description of each step.

4.1   Properties Mining

In this step, the objective is to find the properties sets that are the most shared by the subset of instances extracted from a dataset. Our assumption is that a property often used by several instances of a given type is more important than less often used properties for the same instances. This set will be then used to calculate a completeness value. More formally, let $\mathcal{D}(C, I_C, P)$ be a dataset, where C is the set of classes (e.g. rdf:type like *Actor*, *City*), I_C is the set of instances for categories in C (e.g., *Ben_Affleck* is an instance of the *Actor* class), and $P = \{p_1, p_2, ..., p_n\}$ is the set of properties (e.g. *residence(Person, Place)*). And let $\mathcal{I}'$ be a subset of data (instances) extracted from $\mathcal{D}$ with $\mathcal{I}' \subseteq I_C$. We first initialize $\mathcal{T} = \phi$, $\mathcal{MFP} = \phi$. For each $i \in \mathcal{I}'$ we generate a transaction t. Indeed, each instance i is related to values (either resources or literals) through a set of properties. Therefore, a transaction t_k of an instance i_k is a set of properties such that $t_k \subseteq P$. Transactions generated for all the instances of $\mathcal{I}'$ are then added to the $\mathcal{T}$ set.

Example 2. Taking Table 1, let $\mathcal{I}'$ be a subset of instances such that: $\mathcal{I}' = \{The\_Godfather, Goodfellas, True\_Lies\}$. The set of transaction $\mathcal{T}$ would be:

$$\mathcal{T} = \{\{director, musicComposer\}, \{director, editing\},$$
$$\{director, editing, musicComposer\}\}$$

Table 1. A sample of DBpedia triples and their corresponding transactions

Subject	Predicate	Object
The_Godfather	director	Francis_Ford_Coppola
The_Godfather	musicComposer	Nino_Rota
Goodfellas	director	Martin_Scorsese
Goodfellas	editing	Thelma_Schoonmaker
True_Lies	director	James_Cameron
True_Lies	editing	Conrad_Buff_IV
True_Lies	musicComposer	Brad_Fiedel

Resource	Transaction
The_Godfather	{director, musicComposer}
Goodfellas	{director, editing}
True_Lies	{director, editing, musicComposer}

The objective is then to compute the set of frequent patterns $\mathcal{FP}$ from the transaction vector $\mathcal{T}$.

Definition 1 *(Pattern). Let $\mathcal{T}$ be a set of transactions. A pattern $\hat{P}$ is a sequence of properties shared by one or several transactions t in $\mathcal{T}$.*

For any pattern $\hat{P}$ (e.g. a set of properties), let $T(\hat{P}) = \{t \in \mathcal{T} \mid \hat{P} \subseteq E(t)\}$ be the corresponding set of transactions. $|T(\hat{P})|$ the *support* of $\hat{P}$ (e.g. the number of individuals having all properties of $\hat{P}$). A pattern $\hat{P}$ is frequent if $\frac{1}{|\mathcal{T}|}|T(\hat{P})| \geq \xi$, where ξ is a user-specified threshold.

Example 3. Taking Table 1, let $\hat{P} = \{director, musicComposer\}$ and $\xi = 60\%$. $\hat{P}$ is frequent as its relative support (66.7%) is greater than ξ.

To find all the frequent patterns $\mathcal{FP}$, we used, as we motivated above, the FP-growth itemsets mining algorithm. However, according to the size of the transactions vector, the FP-growth algorithm could generate a very large $\mathcal{FP}$ set. Furthermore, we need only frequent enough patterns. In itemset mining, a concept, called "Maximal" frequent patterns, allow us to find those subsets of properties. Thus, to reduce $\mathcal{FP}$, we generate a subset containing only "Maximal" patterns.

Definition 2 *(MFP). Let $\hat{P}$ be a frequent pattern. $\hat{P}$ is maximal if none of its proper superset is frequent. We define the set of Maximal Frequent Patterns $\mathcal{MFP}$ as:*

$$\mathcal{MFP} = \{\hat{P} \in \mathcal{FP} \mid \forall \hat{P}' \supsetneq \hat{P} : \frac{|T(\hat{P}')|}{|\mathcal{T}|} < \xi\}$$

Example 4. Taking Table 1, let $\xi = 60\%$ and the set of frequent patterns $\mathcal{FP} = \{\{director\}, \{musicComposer\}, \{editing\}, \{director, musicComposer\}$ $\{director, editing\}\}$. The $\mathcal{MFP}$ set would be: $\mathcal{MFP} = \{\{director, musicComposer\}, \{director, editing\}\}$.

4.2   Completeness Calculation

In this step, we carry out for each transaction, a comparison between its corresponding properties and each pattern of the $\mathcal{MFP}$ set (regarding the presence or the absence of the pattern). An average is, therefore, calculated to obtain the completeness of each transaction $t \in \mathcal{T}$. Finally, the completeness of the whole $t \in \mathcal{T}$ will be the average of all the completeness values calculated for each transaction.

Definition 3 *(Completeness). Let $\mathcal{I}'$ a subset of instances, $\mathcal{T}$ the set of transactions constructed from $\mathcal{I}'$, and $\mathcal{MFP}$ a set of maximal frequent pattern. The completeness of $\mathcal{I}'$ corresponds to the completeness of its transaction vector $\mathcal{T}$ obtained by calculating the average of the completeness of $\mathcal{T}$ regarding each pattern in $\mathcal{MFP}$. Therefore, we define the completeness $\mathcal{CP}$ of a subset of instances $\mathcal{I}'$ as follows:*

$$\mathcal{CP}(\mathcal{I}') = \frac{1}{|\mathcal{T}|} \sum_{k=1}^{|\mathcal{T}|} \sum_{j=1}^{|\mathcal{MFP}|} \frac{\delta(E(t_k), \hat{P}_j)}{|\mathcal{MFP}|} \tag{1}$$

such that: $\hat{P}_j \in \mathcal{MFP}$, and $\delta(E(t_k), \hat{P}_j) = \begin{cases} 1 \ if \ \hat{P}_j \subset E(t_k) \\ 0 \ otherwise \end{cases}$

Example 5. Let $\xi = 60\%$. The completeness of the subset of instances in Table 1 regarding $\mathcal{MFP} = \{\{director, musicComposer\}, \{director, editing\}\}$, would be:

$$\mathcal{CP}(\mathcal{I}') = (2 * (1/2) + (2/2))/3 = 0.67$$

This value corresponds to the completeness average value for the whole dataset regarding the inferred patterns in $\mathcal{MFP}$.

5 Experiments

The experiments were performed on the well-known real-world datasets, DBpedia, publicly available on the Linked Open Data (LOD). DBpedia, is a large knowledge base composed of structured information extracted collaboratively from Wikipedia. It describes currently about 14 million things.

For evaluating the completeness of different versions of DBpedia, we chose three relatively distant versions. The first one (v3.6) was generated in March/April 2013, the second one (v2015-04) in February/March 2015 and the third one (v2016-10) in October 2016. For each dataset we have chosen a couple of classes from different natures. We studied the completeness of resources that have as classes the following ones: $C = \{Film, Organisation, Scientist, PopulatedPlace\}$. For the properties used in the resources descriptions, we have chosen the English datasets "mapping-based properties", "instance types" and "labels". The number of triples (statements) of each class is given in Table 2.

Table 2. Number of resources/class

	Film	Organisation	PopulatedPlace	Scientist
v3.6(2013)	53,619	147,889	340,443	9,726
v2015-04	90,060	187,731	455,398	20,301
v2016-10	106,613	275,077	505,557	23,373

In the first step of our experiments, we constructed the set of corresponding transactions $\mathcal{T}$. A transaction vector is constituted of sequences of properties deduced from instances belonging to a single class (e.g. the set of *Film* in DBpedia). The set of transactions is then used as an input to generate frequent patterns and to compute the completeness. Experiments were run on a Dell XPS27 with an Intel Core i7-4770S processor and 16 GB of DDR3 RAM. The execution time of each experiment is about 2 min.

The evaluation methodology consists of calculating, regarding the same inferred schema (for us the same $\mathcal{MFP}$), the completeness of different versions. We chose, as a reference schema, for our experiments the one inferred from the older version. Thus, we can observe the completeness evolution over versions.

We performed a set of experiments on which completeness calculation was performed only on equivalent resources belonging to the three versions (intersection of triples subjects). Therefore we focus on how the completeness is impacted by updated (not inserted) data. Figure 1 shows the completeness results obtained for the chosen classes of DBpedia v3.6, v2015-04 and v2016-10 when varying the minimum support ξ. The completeness is calculated for the three versions regarding the same $\mathcal{MFP}$ inferred from the v3.6 version.

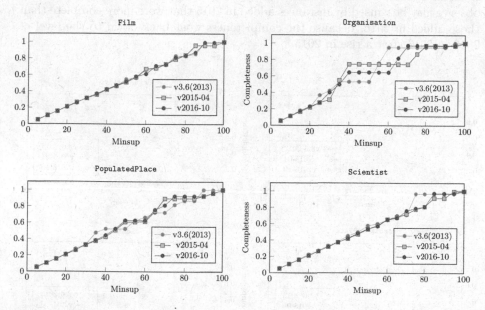

Fig. 1. Completeness of equivalent resources from DBpedia v3.6, v2015-04 and v2016-10

The results show that the diagrams of the classes *Films* and *PopulatedPlace* are roughly the same. Thus, we can conclude that for these classes, either the resources where almost not updated, either the updates preserve the completeness. For the class *Scientist* and *Organisation*, the results show that completeness seems better in 2013 version. This may be due to the addition of new widespread properties across those classes but not added for enough individuals, or maybe some individuals have lost some important properties while DBpedia evolving.

We then reproduced the same experiment but this time by taking all instances of a given class (not only instances belonging to the three datasets). The results of this new experiment are given in Fig. 2.

We observe for *Scientist* and for *Organisation* (except completeness for ξ between 30% and 50%), that completeness values are almost the same. As completeness was better in first experiment in 2013 for those two classes, we may think that added individuals in 2015 and 2016 raised the completeness. However, for *Films* and especially for *PopulatedPlace* there is a clear difference in completeness values, and surprisingly 2013's and 2016's versions are very close. For the class *Films* the completeness values get lower in v2015-04 compared to the v3.6 version and the v2016-10. Either new *Film* instances lack of important properties when added, either updated *Film* instances have loss some of their important properties. Since in the first series of experiments, we see there are no differences for common instances for the three DBpedia versions, we can conclude that the first justification must be the right one. For the *PopulatedPlace* class, as updated instances did not change the completeness, the result we can observe may be caused by instances added in 2015 that were more complete than those added in 2016, because the completeness went back down to the level of 2013 in 2016 after a rise in 2015.

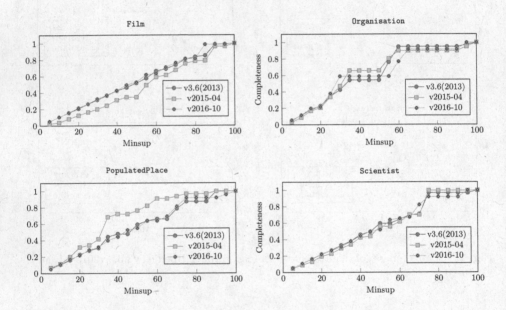

Fig. 2. Completeness of DBpedia v3.6, v2015-04 and v2016-10

6   Conclusion

This article is an exploratory study on the evolution of DBpedia completeness. We have first presented a completeness calculation approach. We then conducted a set of experiments upon three relatively distant versions of DBpedia, and we applied our proposed approach to four classes of DBpedia, any other classes could have been used instead. These experiments revealed that datasets completeness could increase or decrease due to changes made to exiting data or to the new added data. We also noticed that often this evolution does not benefit from the initial data cleaning as the set of properties continue evolving over time. Our approach could be helpful for data source providers to improve, or at least to keep a certain completeness of their datasets over different versions. It could be particularly useful for datasets constructed collaboratively, by imposing to contributors some rules when they update or add new resources. In the future, we plan to enrich our investigation with other data sources such as Yago, IMDB, etc. We would like to study the reasons why some categories improved their completeness over time while other do the opposite. We also plan to improve the proposed approach by randomizing the selection of classes and increasing the number of those selected classes. We are currently working on an approach for Linked Data completeness improvement directed by the dataset content, its completeness results and *owl:sameAs* links.

References

1. Batini, C., Cappiello, C., Francalanci, C., Maurino, A.: Methodologies for data quality assessment and improvement. ACM Comput. Surv. (CSUR) **41**(3), 16 (2009)
2. Eastman, C.M., Jansen, B.J.: Coverage, relevance, and ranking: the impact of query operators on web search engine results. ACM Trans. Inf. Syst. (TOIS) **21**(4), 383–411 (2003)
3. Grahne, G., Zhu, J.: Efficiently using prefix-trees in mining frequent itemsets. In: Proceedings of the ICDM 2003 Workshop on Frequent Itemset Mining Implementations, 19 December 2003, Melbourne, Florida, USA (2003)
4. Han, J., Pei, J., Yin, Y., Mao, R.: Mining frequent patterns without candidate generation: a frequent-pattern tree approach. Data Min. Knowl. Discov. **8**(1), 53–87 (2004)
5. Hartig, O.: Trustworthiness of data on the web. In: Proceedings of the STI Berlin & CSW PhD Workshop. Citeseer (2008)
6. Herzig, D.M., Tran, T.: Heterogeneous web data search using relevance-based on the fly data integration. In: Proceedings of the 21st World Wide Web Conference 2012, WWW 2012, Lyon, France, 16–20 April 2012, pp. 141–150. ACM (2012)
7. Hogan, A., Harth, A., Passant, A., Decker, S., Polleres, A.: Weaving the pedantic web. In: Proceedings of the WWW2010 Workshop on Linked Data on the Web, LDOW 2010, 27 April 2010, Raleigh, USA (2010)
8. Mendes, P.N., Mühleisen, H., Bizer, C.: Sieve: linked data quality assessment and fusion. In: Proceedings of the 2012 Joint EDBT/ICDT Workshops, pp. 116–123. ACM (2012)
9. Pipino, L.L., Lee, Y.W., Wang, R.Y.: Data quality assessment. Commun. ACM **45**(4), 211–218 (2002)
10. Schmachtenberg, M., Bizer, C., Paulheim, H.: Adoption of the linked data best practices in different topical domains. In: Mika, P., et al. (eds.) ISWC 2014. LNCS, vol. 8796, pp. 245–260. Springer, Cham (2014). doi:10.1007/978-3-319-11964-9_16
11. Wang, R.Y., Strong, D.M.: Beyond accuracy: what data quality means to data consumers. J. Manag. Inf. Syst. **12**, 5–33 (1996)
12. Zaveri, A., Rula, A., Maurino, A., Pietrobon, R., Lehmann, J., Auer, S.: Quality assessment for linked data: a survey. Sem. Web **7**(1), 63–93 (2016)

Author Index

Printed in the United States
By Bookmasters